P9-BIE-645

BASIC SYSTEMS ANALYSIS

BASIC
SYSTEMS
ANALYSIS

RICHARD W. LOTT, Bentley College

Canfield Press • San Francisco
A Department of Harper & Row, Publishers
New York, Evanston, London

COVER DESIGN by Doug Luna

Book Design by John Reamer

BASIC SYSTEMS ANALYSIS

Copyright © 1971 by Richard W. Lott

Printed in the United States of
America. All rights reserved. No part
of this book may be used or
reproduced in any manner whatsoever
without written permission except in
the case of brief quotations embodied
in critical articles and reviews.
For information address Harper &
Row, Publishers, Inc., 49 East 33rd
Street, New York, N.Y. 10016

Standard Book Number: 06-385320-5

Library of Congress
Catalog Card Number: 72-143694

PREFACE

Basic Systems Analysis is written for those who hope to learn
techniques and gather ideas to help them design better systems.
Systems experts have not yet agreed upon a standard systems
curriculum. The learning approach which seems to be used with
the most success is presentation of the principles, having the
student apply those principles to structured applications, and
then putting him into real life situations. The end-of-chapter
questions have been made as realistic as possible.

Rather than covering the details of such things as equipment
capability and computer programming, this book concentrates on
objectives and the information that should be processed. That is,
what should be in a report has been given more treatment than
how to go about preparing the report.

Although this book is written from the standpoint of business
systems, the concepts apply to government and nonprofit
organizations as well.

Waltham, Mass.   Richard W. Lott
March, 1971

CONTENTS

13. Barriers and Pitfalls   215

14. Other Comments About the Systems Field   227

15. Additional Exercises and Case Studies   235

Index   257

BASIC SYSTEMS ANALYSIS

CHAPTER 1

Concept of a System

The typical dictionary defines "system" as a number of activities united by some form of regular interaction or interdependence. In a business sense, a system represents all the tasks that must be done in order to provide information required by people who have been given responsibility to run respective portions of the business. The terms "method" and "procedure" will be used interchangeably, and they will refer to the individual activities that are assembled into a system.

Thus, a payroll system may include, but would not be limited to, the procedures of:

1. recording hours worked by employees.
2. developing gross pay.
3. calculating deductions.
4. printing checks and payroll registers.
5. distributing checks.
6. updating individual earnings records.

Filing payroll tax returns is a portion of the task of a payroll system as well. It involves a number of individual procedures which can be collectively referred to as a subsystem. A system, therefore, can be composed of individual procedures, subsystems, or a mixture of both, of necessity closely related to each other.

If you have ever been employed where payroll checks were consistently late or have made purchases at a store that has made billing errors regularly, you may have been inclined to say "they have no system." Actually, whatever means a company uses to get the job done is really a system, but in the two cases above, the systems are poor ones. A sales representative of a computer manufacturer or a computer operations specialist may be referring to the computer or related hardware when he speaks of the "system." However, the use of expensive equipment does not guarantee a good system. When the Houston Astros of the National Baseball League began playing their home games in the $32 million air-conditioned Astrodome, their record did not improve relative to the other teams in the league. In order to have a good system the right people must work closely together.

A typical system in operation today may have grown to its present size and complexion as a result of additions made to an existing system when the need arose. Present payroll systems may have evolved by adding or changing steps to pay people by check rather than cash, by providing for the various fringe benefits and

1

payroll deductions, and handling details related to time off and tax information.

In earlier years, the necessary changes were most likely made by people who worked on a full-time basis in the Payroll Department. As time went on, management probably noticed that the people in the Payroll Department were really too busy operating their present system to consider making changes. Management also may have noticed that work relating to other departments was being done in the Payroll Department, and that much more coordinating was required throughout the organization.

It is at this point that a Systems Department may have been set up by selecting people from regular operating departments such as Payroll, Billing, and Inventory Control. With the absence of a daily routine, those selected to be part of the Systems Department could then concentrate on the planning needed to effect the systems required. The trend indicates that even a small business may establish a Systems Department.

The people who study and design systems are called *systems analysts*. Although this is their title, systems analysts may spend more of their time on design rather than analysis. This book is going to present ideas that will help you to develop your ability to become a business systems analyst.

The following is a list of functions that a systems analyst may perform.

1. Many paper forms are designed on which transactions will be recorded and reports will be prepared. These forms will be used throughout the organization.
2. What information is needed by various employees to adequately perform their jobs is determined.
3. The worth of a new system is estimated by relating improvements in operations to the cost of providing for them.
4. Equipment studies are made in order to recommend the most suitable equipment for the job.
5. People are trained to perform certain jobs. For instance, the people in the Systems Department may be responsible for the decisions to acquire office machines; it may be their job to teach courses that familiarize employees with the use of adding machines and desk calculators.
6. Sequences of manual and automated steps are designed and computer programs are written which will furnish the information needed for effective operations.

If you were to review the current help-wanted advertisements for systems analysts, you would find a most encouraging market, with above-average earnings and room for extensive growth. Also, depending upon your sources of information, you would find the demand is considerably greater than the supply, a situation that probably will continue for many years to come. If entrance into the field and subsequent success were easily attained, thousands of people would fill the vacancies. The salaries would not be so attractive and the jobs would not be so growth-oriented. Since this has not occurred, we can conclude that such a position is not so easily filled. Any success in this profession will come as a result of determination and hard work on your part.

It is not easy to teach someone to become a systems analyst, since he will not always do the same things as he proceeds from one system to another. The things which are common to several systems are not always done in the same order, and the systems analyst will always be working through people who will be substantially different from those he encountered in his previous assignment. In other words, there is no "cookbook" approach, no magic formula that says you do this much of such and such now and then proceed to do that much of something else. The person who does this work well has been presented (formally or informally) with certain basic concepts that would apply in any area. The tools and means are hopefully made available to him by his employer with certain restraints; and then it is up to him to determine what has to be done and how to do it. It is his own creative ability that determines what he accomplishes. As you will soon see, the amount of cooperation the systems analyst gets also has great bearing upon his results.

Purposes and Benefits

The basic purpose of a business system is that of furnishing units of information to the proper people so that they can operate according to a plan. Information is typically needed to:

1. compensate employees for services rendered.
2. order, receive, and pay for products and services obtained from outside parties.
3. collect from customers for products and services they receive from us.
4. anticipate customers needs so they can be served promptly when they place an order.

5. calculate tax liabilities and liquidate them on time.
6. determine profits and distribute them to owners.

For several years much of the discussion about systems analysis has taken up with "management information systems." Such systems are purported to exist or be required so that management has everything it needs to make decisions. You can be assured that not all systems exist to enable management to make decisions, as you can readily see after examining the six examples above.

For example, the steps required to print weekly payroll checks may present very little information to a manager which will enable him to do his job more adequately. This doesn't mean that he isn't concerned about the figures. It just means that there may be no one who is anxiously awaiting information from the report. People all through the organization need information to do their jobs.

1. The shipping clerk needs information so that he knows the number of items to send and to whom they should be sent.
2. The maintenance employee needs information that will enable him to know what machine to fix or which machines to grease on a preventive maintenance program.
3. The credit clerk must get information regarding a prospective customer so that he can make proper investigation of his credit standing.
4. The personnel manager needs information to determine who is the best of ten applicants for a job.
5. The treasurer needs information regarding spending plans to determine how much money must be obtained.

Thus, systems analysis involves the development of procedures that will provide appropriate information for all people in an organization. From a practical point of view, there should be more concentration on what is done rather than how it is done. Every system must be examined to make sure it is serving its purpose in addition to meeting certain benefit/cost criteria. It is possible to determine if operation of a new warehouse will have any impact upon profits. The value of a sales analysis report can be related to the cost of providing it. On the other hand, the benefit derived from providing pay checks, that of retaining employees, can not quite be related to the cost of printing them. In the latter case, you can't eliminate the service but must render it as efficiently and effectively as possible.

At the same time you are making out pay checks you may be providing something that is linked directly with some other important segment of the business, such as developing the cost of making a product. Regarding an application such as sales forecasting, you won't lose all your customers if you don't have a formal forecasting program, but you may not have a satisfactory sales volume if you can't meet customer demand. The point is that you will either incur the costs of providing certain information or suffer the consequences of not providing it. In 1969, a computer manufacturer went bankrupt despite the fact that he was selling a well-built product for which he had a substantial backlog of orders. The company reportedly had no control over purchases, inventory, or costs, and it eventually ran out of cash.

America has often been referred to as the "land of plenty," particularly in regard to our eating habits and our physical possessions. Many people feel that the typical business also has more information on a collective basis than it needs to operate effectively. And the worse part is that it often does not have the things it needs most. Thus, a business could conceivably benefit by eliminating some functions and adding ones that would be more worthwhile. Most of the challenge of systems work is to eliminate that which is not needed and add what is really needed.

At least one reason for the overabundance of information is the desire of a person in any given position for protection—the desire to have on hand any and all things that may be remotely concerned with that position. This may be owing to an individual's own feeling that he is more important if he is involved, to the reprimand he once got from a supervisor for not having certain information, or to the fact that he does not really know what he needs and tries to compensate by having as much information as possible in the hope that a part of it will work. Thus, many people arm themselves with much more than they need or can use effectively.

Another contributor to this unfortunate situation is the growing ease with which data is obtained. With computer printers capable of rated speeds of up to 2500 lines per minute, printed lines 15 inches wide, and the quality of paper and carbon sufficient to give multi-copies, it has almost become too easy to get reports. Couple this with the great boom in the use of office copiers and you can see how easily the average businessman can accumulate much more data than he can digest.

Not only has the businessman too much data, he may have too many people working for him. This may be caused by the desire to hire heavily enough for peaks of activity so as to prevent overtime and consequently having too much manpower in slack times. One of the more difficult things you will ever accomplish is getting a supervisor to volunteer to reduce his staff—or even to listen to anyone's pleas to do so. It is often the supervisor's idea that his salary is related to the number of people he supervises, so he will think carefully about any moves to reduce his staff.

Quite often, though, the efforts of the systems analyst will not cause the immediate reduction of the number of people but may result in fewer people to be hired as volume rises. This may be in keeping with certain management policies of not directly laying off anyone; attrition can often be counted upon to reduce the work force when the number should actually decline. This practice will have a beneficial effect on all workers in the company. It may be better from an overall point of view to keep an inactive person on the payroll for several months rather than fire him outright. Volume may increase or opportunities may arise in the time which may be used effectively. If twenty people spend a few minutes with each other every day grumbling about the bad deal John Doe got, you have soon wasted more money than if John Doe were still drawing his pay.

What Is Left to Do?

You may ask the logical question that if a company has been in business fifty years, then they have obviously had some version of payroll system for fifty years, and therefore, "Why would there be any systems work to do in regard to the Payroll Department?" The following should help to explain why systems development is never really complete.

1. Opinions change regarding needs and how they should be met. A person may change his own mind or a new supervisor may have ideas vastly different from the previous supervisor. A new sales manager is not likely to suggest that the company quit sending invoices to customers, but he may want ten sales reports eliminated and fourteen new ones put into effect.

2. New data processing equipment may make new techniques desirable or possible. Some systems effort must be expended to put the new equipment to work.

3. In many areas of business the larger portions of savings have already been obtained. For instance, some companies feel they have cut the cost of raw material almost to the lowest possible level. Some feel they have automated to the point where not much more can be squeezed from direct labor. For many companies increased profits are now coming from increased volume and not because the profit rate is up. In fact, because of various pressures, many of which the business has little control over, it is getting more difficult to operate at reasonable profit levels; and a company has to press on into new areas to continue to save. General Motors Corporation, regarded as one of the best managed and most efficient businesses, has seen its profit rates decline as follows:

Table 1.1   Net income as a percent of:

	Sales	Total Assets	Net Worth
1964	10.21	16.85	22.83
1965	10.25	18.60	18.51
1966	8.87	14.68	20.55
1967	8.13	12.26	17.57
1968	7.61	12.36	17.75
1969	7.04	11.54	16.72

4. Outsiders influences are making greater demands upon the business. The following are two examples.

(a) Several years ago I worked in a city that adopted a 1 percent income tax on December 22, effective the following January 1. The tax went into effect, with payroll deductions of 1 percent on gross pay. The city recognized that the tax might be illegal, so they did not spend the money they received. The money collected in the first quarter of the year was used to buy interest-bearing bonds; collections after April 1 were deposited and left idle in a checking account. Late in May, the tax was ruled illegal and all the money collected was to be returned to those who had paid, along with interest income on that portion of taxes paid in the first quarter. Some people elected to donate their withheld taxes and/or interest thereon to the city. Late in June, legislation was passed which made a city income tax legal and a ½ percent income tax went into effect July 1. Although the money that the employer had to collect was actually an expense to the employee, not the employer, there was a substantial amount of systems work for employers to try to keep up with the taxing body. Also, all employers are now incurring the costs of collecting the tax, and they do not receive any compensation for their efforts.

(b) In 1969 the United Auto Workers Labor Union obtained a provision in their contract whereby automobile company employees would be reimbursed for drug prescription costs in excess of $2.00. Systems had to be designed to make this program work smoothly.

5. There is a desire for things to happen faster. Although the typical approach to speeding things up has been to get faster equipment, it actually takes vast systems changes to get information where it is needed on a more rapid basis.

6. There is a continuing need to reduce overlap that can so easily occur in any operation. Since many systems have just grown up over many years, there is likely to be a great duplication of efforts.

7. Changes in a company's physical operations require new systems. An airline has had to make drastic changes in its method of operation as it has proceeded from a DC-3 to a DC-6 to a Boeing 707 to a Boeing 747 fleet of planes.

8. Failing to properly plan for substantial increases in the volume of transactions has produced great problem areas.

(a) Service at the Chicago Post Office once broke down completely, and it took three weeks to clear out the backlog.

(b) A number of telephone companies have insufficient lines available for calls and too few people to handle service requirements.

(c) Electric companies in many cities have had to ration electricity on hot days when electricity requirements are at a peak. Some electric companies have eliminated their advertisements for air conditioners because they would not be able to supply the new users with adequate power.

(d) Stock exchanges have had to close one day of the week or have a shorter working day in the hope of cutting volume enough to provide brokerage firms with time to catch up on their paperwork. In the past, volume increases were handled by hiring more people and getting more space, but by the late 1960's that approach was no longer feasible.

So you can see why the systems approach to business operation has become so important and so intense. But the work suggested by the eight examples above does not economically lend itself to redesigning every system each year or so. Obviously a company could incur large losses as a result of spending too much money on systems work, just as it could by operating too many ineffective systems.

Two thoughts that apply at this point are these.

1. Why do we always have time to do something over that we didn't have time to do right the first time?
2. Nothing is so futile as doing that with great efficiency something that shouldn't be done at all.

From these two statements you should understand that the systems analyst's job is to design a system once and design it correctly if it must be designed at all, then redesign it only when conditions truly warrant.

Dynamic Change in Business Methods

A short time ago someone compressed an estimated 50,000 years of the existence of man into a period of fifty years and thereby developed the following chronological order of events.

For nearly forty years, man lived by hunting wild animals and sleeping in a cave.

The first city was built twelve years ago.

Writing was developed seven years ago.

Christianity appeared two years ago.

Guttenberg invented the printing press six months ago.

Edison demonstrated the light bulb one month ago.

The Wright Brothers flew their airplane for the first time three weeks ago.

The atomic bomb and the first jet airplane appeared one week ago and commercial television followed shortly after that.

Sputnik I was orbited by the Russians two days ago.

The first manned space capsule was put into operation last night.

It is clear that the pace of change is increasing and businesses must be ready to do what is necessary to meet such challenges. Here are examples of some things that business is handling differently in comparison to methods used previously.

1. Companies no longer leave large excess cash balances in noninterest earning accounts, even for one day. Westinghouse Electric Corporation has all of its offices "wire" excess funds to its home office every day. From there the funds are loaned out until the next day. Such a practice on the part of depositors has caused banks to be more aggressive in obtaining loanable funds.

2. A new bank trying to rapidly build up its checking account business offered unlimited, free activity to anyone who would maintain a minimum of $100 in his account. This caused most of the other banks to adopt a similar program at a loss of service charge income to them.

3. Selling companies are establishing minimum order quantities in order to eliminate some of the inefficiency of processing small orders. A company can actually lose money on every $5 order it handles.

4. Many businesses are not so hesitant about getting rid of unprofitable items. International Business Machines Corporation does not hesitate to eliminate a product that can't be made to return at a specified profit rate.

5. There is more cooperation among certain businesses as they see an erosion of their profit margins. The major airlines have entered into a pact where extra profits earned because of a strike at a member airline are turned over to the airline that is struck by its employees.

6. At one time, automobile companies used as a defense on a defective car suit the idea that since they had sold the car to a dealer, they then weren't liable to the individual who bought the car. Automobile manufacturers now accept a much greater responsibility for their defects by notifying owners of all cars thought to be defective and offering to fix the defect free of charge. One manufacturer sent 5,000,000 letters to customers in one year telling them to take their car to a dealer for checking and possible free service.

7. Fountain pens have been replaced by ball-point pens and felt-tip markers; 78 r.p.m. phonograph records have been replaced by long-play albums.

Changes in the basic way of operating a business have required new systems.

Who Is Involved in a System?

Everyone in an organization is either directly involved with a system or is in some way affected by it. Obviously, each person is involved with the payroll system as it relates to the things that he must do in order to record his time on the job. Each individual is also concerned with the correctness of his pay and payroll record as well as being paid on time.

Review any other system and trace through from the origin of raw data until a report is prepared or a file is updated. The people who are directly involved are probably easy to spot, and the operation probably includes only a small percentage of the total work force. But every employee is affected by any system that is so poor that profits suffer—pay rates thus may not be as high as might otherwise be possible; fringe benefits and working conditions may not be up to par; possibilities of internal growth may not be present. Furthermore, morale suffers when adverse publicity prevails about a particularly bad company situation. So systems are everyone's business, and no one can afford to do other than work for maximum improvement.

Employee involvement must start at the very top and work its way down. In fact, more systems failures appear to be caused by the lack of top management involvement than by any other cause. Perhaps the reason for this is managers so often feel they have discharged their responsibility to the systems effort when they have signed the contract that orders equipment and okayed next year's budget. But successful systems don't develop this way. The systems specialist is not necessarily a person who knows what is needed to run the company. To allow a person inexperienced in any phase of management to determine what management will have available to work with will probably create chaos.

Management may make the mistake of thinking that a series of bad breaks is at the root of many of their troubles and not poor systems. Avis Rent-A-Car keeps its management actively involved in what is going on by having people such as the president and the comptroller spend at least one week each year performing tasks such as writing up car rental orders and handling service requests.

So top management must be concerned with requirements of the system in addition to providing the necessary resources. And as the systems study gets underway, management must see that employees are made aware generally of what is happening and the types of changes that can occur as well as the expected results. Few things will undermine employee confidence sooner than an attitude that many things are being done behind their backs. Give the employees an idea what is going on, over the president's or general manager's signature, and solicit their cooperation and emphasize that it is necessary and expected.

In the book *The Desk Set* (later produced as a Spencer Tracy-Katherine Hepburn movie), a consultant dropped in from nowhere and began disrupting business activities in preparing for a computer. Not a word was heard from management until the day the computer first printed pay checks and inserted a pink slip into every employee's pay envelope. Only then did management disclose the fact that the computer was being obtained to take over some of the tedious jobs so people could be freed to do more work appropriate to their skills.

An example of an excellent way to keep people informed is shown in Figure 1-1.

Top management must also make decisions that heavily bear upon the restraints of the system. For instance, the systems group cannot be allowed to run wild doing things that will greatly disrupt the organization. If an airline is about to install a real time system to process seat reservations, top management must be aware of any consequences of machine downtime. If you were the president of an airline, wouldn't it come as a shock to you to find that people calling in for tickets were told to call again in three hours with the hope that the machine would be operational by then? The president should have been involved in earlier design stages where such risks were exposed and alternatives discussed. Of course, the typical solution to this has been to have a duplicate machine on hand for backup. And the duplicate machine need not necessarily sit idle for the purpose of being ready when the other fails; the backup machine can be doing regular batch work until it gets a signal that it is to switch itself to real time transactions.

Additional management participation is also required to check on the progress of the systems analysis staff. Because these people represent such an important activity and may be difficult to hire is no reason why they should be granted undue privileges. After all, there is no one anywhere who does not have some person or some group bigger than he is. Even the president of a large corporation or the President of the United States has responsibilities to superiors. This is not in any way an endorsement of the Big Brother approach, merely an indication that every person and group must answer to someone above.

Importance of People

It cannot be emphasized too strongly that systems must be designed for use by people, although in some cases it appears that everything has been done to satisfy the requirements of a machine. Some companies have realized how they have alienated employees, custom-

Figure 1-1.   A letter informing employees about plans to get a computer. (Reprinted with permission from National Industrial Conference Board Report No. 98, *Administration of Electronic Data Processing*, 1961, p. 119.)

THE HEAD OFFICE OF THE

SUN LIFE ASSURANCE COMPANY OF CANADA

MONTREAL

ANNOUNCEMENT TO HEAD OFFICE STAFF

On May 17th a card-controlled Electronic Machine of medium size, will be installed in our Head Office. This machine, the first in Montreal, will be of assistance to us in carrying on our normal work through the Hollerith Department. I understand, for example, that it will enable us to handle, in one process, work which now requires two or three separate processes on less comprehensive machines.

At this same time I am pleased to announce that the Company has entered into arrangements with (a manufacturer) to have a large magnetic tape-controlled electronic computer, called a (model), installed in our office early in 1957. This we expect will be the first (make) installed at any point outside the United States. It will be the first in Canada.

It is expected that the system will enable the Company's work to be performed more economically and more efficiently than we have been able to do heretofore. We believe that much work of a monotonous, repetitive nature will in due course be handled by the machine, thus allowing members of our staff to devote themselves to that work which does require intelligent understanding of the basic transaction being handled. We believe that it will result in making many jobs done by our staff more interesting and rewarding.

I want to assure all of you who are now on the staff of the Company that although the changes in our work which will result from using this modern equipment may affect the work which you are now doing, it will not result in the downgrading of any individual or the release from our service of any individual. On the contrary, I believe that in due course this is going to result in the upgrading of many posts held by our employees. These jobs will become broader in concept and in the responsibility carried. In due course many of you will be given the opportunity of qualifying for new jobs.

This new equipment will enable the Company to continue to maintain its place as one of the great life insurance companies of the world, continuing to provide an economical, effective, expanding insurance service to the public of the countries in which we operate.

President.

ers, and vendors by such an approach, and many have placed prominent advertisements indicating how they have attempted to revert systems to the people approach. For several years Hertz and Avis centered their advertisements around their companies being

"No. 1" and "No. 2," while National Rent-A-Car was placing the customer first in its ads. There is nothing basically wrong with using machines, but you must design systems so that people and machines are properly blended to obtain the best results from both. In the final analysis, it will be people that make the system work.

Not only must you recognize the existence of people as a collective group, you must also realize the vast differences in people, one to another. A baseball manager must learn that a pat on the back is the basic type of recognition that one player expects but that the only language another understands is a bigger contract. So also, the systems analyst must learn how to treat the individuals he serves. Perhaps a good general practice is to follow the Golden Rule.

One company that bought an additional firm in California decided to consolidate some of the control activities and determined that a purchasing agent at the California location should be moved to the home office of the surviving company on the East Coast. The person chosen for the job was sixty-three years old and did not want to move 3,000 miles to a position that would last only the two years until he would retire. The news of this forced change in locations soon traveled to the home office of the company. Then countless man hours were spent by people discussing how inconsiderate management was for requiring this person to relocate so near to his retirement. The company just had not properly considered that move.

As soon as there is more than one person in an area, there is bound to be a growth of "office politics." This represents those activities perhaps above and beyond the call of duty which a person will resort to, normally to enhance his own self-interest. Thus, politics may sometimes get in the way of doing things objectively, and you must learn to put these things in their proper perspective. A friend of mine once flew to another city for a job interview. During the day in the company's offices, he was instructed on how to fill out a form for his travel expenses. When he remarked on how liberal the travel reimbursement policy was, he was told how happy everyone in the company was with it and should he join them, it was with the understanding that this was one area he wouldn't investigate too closely. Another company had the policy of automatically deducting $1.00 from every bill it paid, over and above any trade discounts or cash discounts it was entitled to. So you can see, you may even have to examine some of your own ethics to determine if you want to be part of such situations.

Which Systems Need Attention?

Since the typical organization will not have as many systems specialists as it could probably use, the Systems Department will have to develop ways of determining the priority areas in which to work. The following suggest prime candidates for study.

1. Any internal area that is constantly complaining that it is not getting what it needs or is not getting it on time should be studied. In following this approach too closely, one could devote all attention to the "wheel that squeaks the loudest" and therefore may overlook the non-complainer who really needs help. Management must see that other areas are covered, too.

2. Everyone in the organization must be alert to recognizing and reporting such danger signals as customer or vendor dissatisfaction. Quite often the reason for dissatisfaction results from a high error rate which causes steps to be corrected and deadlines to be missed. Some possible solutions to the error rate are making the work more interesting, using better training programs, developing different methods of determining employee aptitudes for certain types of work, and automating repetitive tasks.

3. A department that has unduly high employee turnover should be studied. Management should establish realistic turnover standards in line with the type of people expected to perform that type of work. Then they can judge actual performance. Perhaps you can even develop such a satisfactory place to work that female high school graduates will not consider it just a short stopping-off place on the way to getting married. Considering the high costs of obtaining and training help, this can be a significant area in which to cut costs.

4. A department that has tremendous swings in volume during the week, month, or year must be studied for maximum utilization of personnel. Perhaps one area has its peak in the first two weeks of the month and another department during the last two. The manager of each may feel he is fully justified in staffing for the peaks. But if top management can see what is happening to costs, perhaps something can be done to mold the two staffs closer together. Notice that top management is required here in order to get the proper authority to cross organizational lines. You may get opposition to a consolidation of this type on the basis the two areas are doing completely different things. Certainly each area has its specialists—but each has generalists who are performing such tasks as typing, filing, and checking. At least look to see if any

blending can take place rather than backing off from the start by saying it can't be done.

5. An area in which actual costs are continually out of line with the budget will require a change. Either the budget is not realistic or people are not following the plan used to arrive at the budget.

6. A category of work in which a specialist spends an unreasonably low percentage of time doing what he was hired to do is inefficient. If the typical salesman spends only 15 percent of the day selling, or if a repairman spends only 60 percent of his time repairing, the system should be reviewed to see if those percentages can be raised.

A common occurrence that necessitates a deviation from an organized approach to realistic priorities is the practice of "firefighting." When the system being used has broken down, something is not happening properly because of insufficient design, and the analyst is called back to patch it up. This can become so prevalent that the analyst is always "putting out fires" and unable to devote his efforts to new planning. It would be nice if the need to put out fires could be eliminated, but for many it seems to be a way of life. It may be possible to partially overcome this by relegating certain personnel to maintaining present systems so others may devote their time to the major revisions and the new programs. Many companies put new employees on such maintenance work so they can learn as much about the company as possible in a short time. It is only natural that a person would want to spend his time on the new and exotic rather than on maintenance, and good people often leave a position because they can't move out of the maintenance work quickly enough. You can see the obvious benefits of getting away from the firefighting approach.

A company must always look for an application of a system which has a steady increase in volume. It may be very easy to handle that by just adding people and work space, but perhaps the better solution is to redesign the basic way the job is done. In the 1950's, the banks found they could not continue to add people to handle checking accounts, and they developed the magnetic-ink processing of checks largely centered about computers. In the late 1960's, the brokerage industry seemed to reach an impasse in handling all paperwork in a manual way, but they had no alternative. So the stock market was closed each Wednesday in a deliberate move to reduce volume.

Samples of Interesting Systems

Please refer to Figure 1-2 to see examples of several unusual systems which are being used. Perhaps you will never work on systems like these, but, on the other hand, you cannot predict what systems requirements will be like in twenty or thirty years. The purpose here is to show you what some people have done to serve the organization's needs by applying some very effective, creative thinking. Several of the approaches cited have been adopted by other businesses as soon as their worth was proved by the originator.

Note that the concepts in the first two examples could be used in rather small businesses, while the other two are more appropriate to large companies.

Questions

1. About how much does the use of expensive equipment have to do with making a successful system?

2. What has caused companies to establish systems departments (rather than letting people in operating departments design their own systems)?

3. Why is it difficult to teach someone to become a systems analyst?

4. List several things that must be done in a payroll system which would not furnish anything for management decisions. List several things that might.

5. How could a company with a very high sales volume lose money and possibly go bankrupt?

6. Why wouldn't it be natural for a manager to try to keep his expenses to the minimum and have as few people working for him as possible? What could top management do to try to instill such good habits in its lower levels of management?

7. What is the likelihood that a systems analyst will be released once he has completed the assignment he is working on? (Assume he has done a good job.)

8. Why don't companies that have been in business a long time have systems so effective that they no longer need systems analysts?

Figure 1-2. Examples of interesting systems.

Business Involved	Nature of Operation	Benefit to Business That Installed	Benefit to Recipients of Service
Grocery Store in France	Grocer has only one sample of each nonperishable item on display. Shopper pulls tab card to indicate selection and picks perishable items in usual fashion while the balance of order is filled by store personnel.	Lowers cost of stocking the store.	Cut in costs of groceries is 25%.
Kaiser Aluminum	Sends blank check to selected vendors. Vendor fills in amount, deposits check in bank, and fills the order.	Saves on cost of the major steps involved in purchasing materials.	Eliminates the need for billing and money tied up in accounts receivable.
Association of American Railroads	Automatic eye system records identification of passing freight cars and makes that data available to anyone who has need for it.	Eliminate visual spotting, manual recording, and related errors and time delays.	Can obtain data regarding car location by inquiry directly from railroad's computer files.
Ticketron, Inc.	Installation of on-line printing devices in various business places to prepare tickets on demand to dozens of events.	Can give better data on seats still available and spread out the marketing effort.	Can get tickets in local neighborhood at grocery stores, drug stores, etc.

9. What seems to be the general trend of profit levels? Why is this?

10. Is it possible that a company could spend too much effort on systems work? Why?

11. As a company grows, why can't management handle that growth easily by hiring proportionately more people?

12. Why would a bank give completely free use of a checking account to anyone who would maintain a $100 minimum balance? Assuming it costs a bank 3 cents to process each customer's check, about how many checks per month would be the break-even point if the bank earns 6 percent on its investment?

13. What is probably the largest cause of systems failures? What can be done about it?

14. What is the most important ingredient in any system? (One word, please.)

15. From the following list, select the system that should probably receive attention first; last.

System	Problem
Invoicing	Invoices seem to be about five days late going out.
Payroll	The average payroll check costs 39 cents to prepare. The competitor's is 33 cents.
Quality Control	Six percent of all shipments are returned because of faulty construction. Warranty costs are up 12 percent from last year.
Shipping	For an item that is in stock, it takes an average of eleven days to get it shipped.

16. This chapter pointed out examples of unusual solutions to systems problems. For each of the following, try to be as creative as possible in thinking up a solution to the problem.

   a. A company in the lumber business finds it takes eight hours to unload a large barge of logs with the use of a crane. Suggest a way to unload the barge by the quickest method possible.

   b. A large grower of tomatoes could not hire enough people to pick them, so he developed a machine for this purpose. But the machine was not very sophisticated and it picked all of the tomatoes regardless of how ripe they were. What two aspects of the problem might the farmer concentrate on in order to find the best solution?

c. A soft drink manufacturer has a great deal of trouble getting his customers to return bottles on time and it requires a great deal of bookkeeping to keep all related records. What aspect of the problem should he concentrate on?

d. A very popular pub in a rainy climate received many complaints from customers that it was uncomfortable standing outside waiting to pay the cover charge. What reasonable step may the pub management take to better serve its customers?

e. Review the sketch below. About 10,000 cars make the daily trip to the mainland. Suggest a very simple change which would have substantial benefits to both the Bridge Authority and its customers.

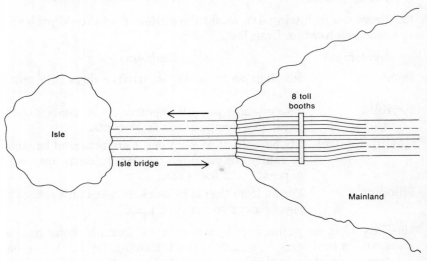

f. During a very hot, dry spell everyone in a city wanted to water his lawn. The city reservoir had plenty of water, but when everyone tried to use it at the same time, pressure in the lines fell to the point where users could not get more than a trickle. Suggest a sound approach to solve the problem.

g. In 1969 a controversy arose as to the best way to show the boundary between the United States and Canada in wooded areas. One group suggested cutting down trees in a strip 20 feet wide. Another group wanted to spray a 20-foot strip of trees so they would lose all their leaves. Recommend another approach that may be nondestructive.

h. A person calling the reservation office of an airline kept getting a busy signal on the telephone. He eventually called another airline that served the same city and was able to make travel arrangements with them. What reasonably simple change might the first airline put into effect to better serve its potential customers?

Other Readings

Brandenburg, Walter H., "Dynamic System Design for Airlines," *Datamation Magazine*, March 1969, p. 26. The author feels that the airlines have concentrated too much on data processing and not enough on moving people and things.

Head, Robert V., "Topless Management Information Systems," *Systems and Procedures Journal*, November-December 1967, p. 28. A "topless" system is described as one that management has shown no direct interest in.

Lott, Richard W., *Basic Data Processing*, 2nd ed. Englewood Cliffs, N.J.: Prentice-Hall, Inc., 1971. Chapter 2 gives an introduction to the topic of systems analysis and design.

Newman, William H., and James P. Logan, *Business Policies and Central Management*, 5th ed. Cincinnati, Ohio: South-Western Publishing Co., 1965. Chapter 2 emphasizes the dynamic nature of business activity and the need to be able to adjust to that change.

Parkinson, C. Northcote, "Our Overheads Are Overhigh," *Management Accounting*, November 1969, p. 9. The author of Parkinson's Law points out that staffing for the highest level of activity obviously causes great periods of idleness; when the peaks return, people cannot automatically get back to the faster routine required. He suggests staffing for the low activity periods and hiring part-time people to handle the peak periods.

CHAPTER 2

Systems Objectives

When a group of people go about organizing a business, they do so with some return in mind. The same holds true of anyone who buys into the ownership of an existing company. The reason for wanting to become an owner is normally referred to as "the profit motive." When a person makes an investment, he does so with a particular investment objective in mind. He accepts high risk with the hope of high returns or accepts considerably less risk with proportionately less returns. Since a company is merely a collection of the ideas of individuals, a company has certain profit objectives. It is not good enough for management to say that the company should make as much money as it can, since this is a rather nebulous type of expectation. It is much better to state in precise terms what is expected. This gives the workers a goal to aim for and also provides management with a better guide for evaluating performance.

When two large corporations merged in 1968, the chairman of the resulting company specified that it should increase its return on net worth by 15 percent and its earnings per share by 10 percent annually. As soon as it was seen that such results were not forthcoming, some executives were fired and others were retired early. There is no getting around the fact that management is going to evaluate the performance of each and every employee.

The problems involved in setting realistic work objectives can be partially alleviated by having the operating people participate in establishing goals. Your first impression may be that those people would set their goals unrealistically low so as to avoid overwork. However, you are likely to be surprised how pride will project itself into meaningful standards. This is particularly true if people can see what their jobs mean in terms of the total picture. Employees will care about their jobs when they can see that management cares enough to keep them informed. This does not mean that the employee should have complete control over the matter; it merely means he should be entitled to have some negotiating power over those things that directly involve him.

Source of Profits

A business exists for the purpose of making money. History is full of cases where companies have ceased to exist almost immediately because profits ceased or never came about. This does not mean that a company will sacrifice everything else for its own gains, but most of the things it does will be weighed in terms of the immediate or long-range effect on the income statement.)

In the hope of maximizing profits, a company may follow practices such as these.

1. Sales can be maximized by having complete lines of products and dealing over wide areas.
2. By internal expansion or by acquisition of going concerns, companies may expand into completely new fields. Thus, one product line of the business may keep going full-scale when there is a dip in the activity of another. The old-time delivery man always relied upon his coal business in the winter when customer needs for ice diminished. Airlines have tried to obtain permission to fly routes that stabilize their business around the year; a New York–Miami route is heavily traveled in the first quarter of the year, just when the New York–London route is not.
3. Companies may attain a certain share of the market.
4. The actual results relating to each product, department, and physical location is analyzed to make sure that each is contributing to what management had in mind when it was established. Of course, the proper thing must be done with those segments that aren't contributing. It doesn't do much good to find out what is wrong if no attempt is made to correct it.

The first three approaches above may help contribute to size and volume but too many companies have found there is not always a direct relationship between total sales and profit. The following figures show how the operating revenue of Allegheny Airlines climbed sharply in its last five years prior to acquisition of another line.

Year	Operating Revenue	Net Income
1963	$24,952,000	$ 302,000
1964	27,875,000	203,000
1965	33,277,000	1,308,000
1966	43,376,000	1,042,000
1967	52,329,000	(516,000) Loss

As you can see, a business that is sales oriented is not necessarily profit oriented. Years ago, movie actor Bing Crosby determined his net take-home pay to be about the same whether he made four movies a year or just one.

An organization such as a railroad may expend efforts on trying to increase traffic or to cut certain costs. The railroad industry has about 1,800,000 freight cars; the average car is in motion from origin to destination only 11 percent of its life. If some means could be developed to double that rate, a railroad could do without a great portion of its fleet or it could substantially increase volume without the need to obtain many more freight cars.

Suppose a railroad has $25 million worth of freight cars and is depreciating them on a straight line basis over 25 years. They could reduce depreciation expense by $500,000 a year ($25,000,000 $\div$ 25 $\times$ 50%) if they could double the effectiveness of their fleet.

It was once considered heresy to do anything that would *lower* sales. It is now common to eliminate a product or a whole division if it doesn't reach a specified goal. Between 1951 and 1960, the Glidden Company (now a division of SCM Corporation) disposed of operations which had sales of $75 million but contributed profits of only about $1 million. Prior to those dispositions the firm had a period of stagnated profits; since then profits have risen consistently.

It is therefore necessary to take a look at each segment of operation, report the facts to management, let them know what it is costing them, and present alternative choices of action along with projected benefits. Hopefully, management will then make wise decisions as to future action.

Systems Requirements

Once profit or volume goals have been established, it is necessary they be broken down and interpreted in terms of information and report requirements. In order to help ensure that profits will be made in the manner required, certain information must be made available to the proper people so they can control the physical operations of buying, manufacturing, and selling. This information is usually submitted in the form of reports that can be rendered to owners, employees, customers, and vendors.

Just what is needed to run a business? The answer to this question must come from many minds. Top management knows (or should know) what they want to keep track of and what they want to be alerted to. They also know the types of decisions they need to make. Middle management often has direct responsibility for the great bulk of workers and one of their great needs is data with respect to employee performance. Since so much of the reporting to top management should normally be in summary or exception form, middle management is often involved in the summarizing and interpreting process. Even the lowest persons on the organizational ladder have needs for information of a very specific but voluminous nature. Since not all information is destined for one organizational level, no one level should be specifying the data that has to be gathered and the information that has to be disseminated. Top management can't afford the time it would take to know all details concerning the clerks' needs, and the clerks obviously aren't in the position of specifying management needs. In a large professional organization, members were billed once a year for annual dues. Only one bill would be sent out and anyone not paying within 30 days would be automatically dropped from the roster. Investigation showed that management of the organization did not really want people to be dropped so easily but because of the relative power of the Bookkeeping Department they went along with the procedure for many years. Thus, various levels must specify what the systems requirements are.

What role should the systems analyst play in determining information needs? Ideally, he should do no more than gather data and eventually satisfy the needs, as reported to him, of the people to be served. Historically speaking, the systems analyst has often determined the whole array of needed information himself. In one company that was getting a computer within two months, the manager of Inventory Control told me he was getting anxious to see what the Computer Department would be furnishing him to work with. Can you imagine the effectiveness of that forthcoming operation if data processing personnel wrote all the specifications without consulting the users? Situations of this type have occurred for some of the following reasons.

1. Management has often taken a hands-off position at a crucial point in systems design. Thus, no direction is given and internal cooperation may not even be requested or mentioned as being required.

2. Personnel in operating departments are often looked upon as a group whose only function is making someone else's system work. It is felt that they need not be allowed to contribute to the design of that system.

3. Systems are too often designed with the preconceived idea that a certain method or machine will be used. Thus, cer-

tain requirements are often rejected because a "275X doesn't do that." Since computer specialists know what a 275X can do, they decide what it will do for their company.

4. Too many systems are justified by the possibility of their cutting clerical costs rather than on the basis of furnishing better tools (information) to work with.

Certainly it is desirable for the knowledgeable analyst to make suggestions that would be helpful or to report worthwhile things in operation in other companies. But to let him completely determine the tools that people shall have to use is too much like having the tail wag the dog. The situation clearly calls for all concerned to work closely together with the systems analyst functioning as a coordinator.

It was previously stated that a major reason for system failure is lack of management participation. A second major reason is the lack of solid problem definition, or too often, no real definition at all. There have been countless cases where a new system is merely a rehash of the old one, using a machine for some functions that had been performed by a lesser machine or by a person. One company previously used tabulating equipment to summarize, for internal purposes, six months' sales to every customer by major product group. The report was turned over to the Marketing Department, which hand-posted the information to a customer ledger history card. The Marketing Department then duplicated the ledger card and sent a copy to the salesman concerned. For any of his customers, the salesman could presumably compare sales in any product group to that of any previous six-month period. But throughout this brief process of redesign, no salesman was consulted regarding what he needed to know about sales to his customers. Furthermore, no attempt was made to inform the salesman as to which products were the most profitable for him to handle, because it was company policy that the salesman was not allowed to know the internal manufacturing cost or the purchase cost of any products entrusted to him to sell. While I am not in favor of broadcasting information about such costs, I think the company might have at least coded these items so that the salesman would know those which were most profitable, those which were least profitable, and those on which the company may actually be suffering a gross loss.

The system was "redesigned" to the extent that a computer replaced certain pieces of tabulating equipment, and the salesman got his six months' report a day sooner than he did before. It was never determined what use a salesman was to make of the report.

It was previously mentioned that reports which are typical of payrolls, accounts payable checks, and bills to customers may be necessary to satisfy certain business requirements but may not furnish much from which to make decisions. Just what types of decisions must be made within a company, and not necessarily by "management"? A partial list follows.

1. Should an existing product be dropped?
2. Should prices be changed, and if so, by how much?
3. Should a new product be added?
4. Should selling terms be changed? Can we afford to continue offering cash discounts in order to induce customers' early payment?
5. Should more credit be extended to a certain customer?
6. Should a new sales office, warehouse, or factory be opened and if so, where?
7. Who should be hired? Who should be promoted?
8. Should certain functions be centralized? Should some be decentralized?
9. Should a piece of equipment be replaced?
10. Should we manufacture or buy a certain product that we are going to sell?
11. How much of a dividend should be paid, if any?
12. Which method should be used to obtain needed money? Should it be a short-term debt, a long-term debt, or sale of common or preferred stock? How much money is needed?

It can be readily seen that information of a specified nature must flow to the proper people so that they can make the best decisions.

Inventory Control

So that you can see how the points in this book can be applied, certain chapters will contain practical use of the principles as related to an inventory control system. Please understand that, in practice, inventory control systems often require several man-years to design, and the material covered herein is not nearly sufficient to be used as the basis for a complete system nor will you emerge as an expert analyst based upon this brief coverage.

(The term "inventory" will generally refer to those finished items, either manufactured or purchased, which are held for resale.) In order to keep this book to a reasonable length, many details you would encounter in practice will be bypassed. We will not normally be concerned with all the details of how people use information, but more with effective ways of getting the right information to them.

Perhaps the ideal situation with respect to inventory would be to have none at all. Here are some possible alternatives.

1. Material could be shipped directly from the vendor to the customer or if the item is sold by the manufacturer, it could be shipped directly from the production line to the customer.
2. If the above is impossible, the inventory could be reshipped to the customer or he could pick it up on the same day it is received from the vendor.

Some businesses have been able to get along with the above arrangements, but most can't effectively operate that way for the following reasons.

1. You must normally buy in large quantities to take advantage of volume and freight discounts. These aren't usually the same volumes in which you sell to your customers.
2. If it is your own manufactured item, you may have to manufacture it in cycles to properly use manufacturing facilities. These are not necessarily the same cycles in which customers buy.
3. Customers often want to be able to see the merchandise they are getting. Many sales at the retail level are made because the buyer can visually compare competing brands. Buying through a catalog appeals to only certain people and is applicable to only certain products.

Discussion of inventory herein will relate to a situation where all inventory is purchased and there is a fluctuating demand for numerous products. If you were a systems analyst in such a company, you would be responsible for developing systems that would help people to control that inventory.

Broad Objectives

If the systems analyst should become involved in a study of inventory control, and notice he is studying inventory control and not just the Inventory Control Department, he should first find out what the objectives of inventory control are. This is often very broadly stated to be "having the right item in the right quantity in the right place at the right time." Such a statement is much too vague to give all the background needed, but note it does give us the following general guide.

1. Management must state what business they are in. They must indicate, for instance, whether the company is in the "pen" business or in the "writing instrument" business, since the latter implies a much broader line of products. It is also necessary to know the general price and quality range in which the business is to sell. Then the inventory analyst can determine what items are to be handled and any which may be sold but not carried in stock. Obviously, it is not possible to handle or stock all the items appropriate to the line of business.

2. The inventory control people must not only know what customers want but also how much they need and how much they can be induced to buy. Naturally, sufficient quantities of all things likely to be sold can't be stocked. This would require a fantastic investment in inventory.

3. There may be various places where the inventory can be stored—in a main warehouse at the home office, in many outlying points, or maybe even in the customers' locations. In any event, we must be ready to provide transportation and warehousing services.

4. Timing in regard to the company's procurement is important because of seasonal variations and normal fluctuations in the demands of customers.

Further investigation shows that top managers, in consultation with operating specialists from all involved departments, have determined the following in a precise way.

1. Inventory must have an average turnover at least five times a year. Turnover is a ratio that measures what was sold in relation to average inventory. For any individual item, turnover would be calculated by dividing the average inventory into the number of items sold. For example, if you sold 320 size A bolts during the year and had an average inventory of 40, turnover would have been

8. Another way of stating this is that the inventory was moved in and out eight times during the year.

Turnover for a group of items or for the whole inventory may be determined by dividing the average inventory (in dollars) into the cost of goods sold (in dollars) as shown on the income statement. If total cost of goods sold was $2.4 million and average inventory was $600,000, turnover was 4. Turnover will tend to fluctuate according to the nature of the business; a banana store would have a much higher turnover than a jewelry store. Turnover will also be a result of the efficiency and timing with which inventory is purchased and sold.

For many people, "turnover" is a very abstract thing. It can be made more meaningful to them to quote "months supply on hand" instead. If turnover occurs four times a year, then months' supply is 3; if turnover is six times a year, then months' supply is 2; etc.

Turnover statistics can be very revealing. If a company is producing a standard item and placing it in a warehouse for sale and if sales are so slow that turnover is only 3 (or months' supply on hand is 4), everyone should recognize that they must be doing something wrong. Over a period of time, a poor or good trend can be determined, and it is also possible to compare your own statistics to those of other companies in the same industry.

2. Total dollar value of inventory should not exceed 28 percent of assets. Based upon budgets, you should be able to determine what the maximum value should be at any time. The actual percentage can be easily checked each time a balance sheet is prepared.

3. Stock should never exceed more than one year's supply of any item. If the annual sales of an item is 150, then inventory should never be allowed to exceed 150. Notice that is a very high level of inventory, and it can be permitted to happen on only a few items or turnover standards could not possibly be met.

4. At least $10 gross profit or a 40 percent gross profit, whichever is larger, should be attained on any sales transaction. If something for $85 were sold which cost $65, a gross profit of $20 would have been made. But the gross profit percentage would be only 24 percent (gross profit of $20 divided by sales value of $85), and that sale would not have met the criteria. A reason for a restriction such as this is the maintaining of proper relationships among all income statement accounts, since the net profit rate won't be satisfactory if the gross profit rate isn't. The typical company probably does not

consider gross profit when it is talking about inventory control. But it is hardly worthwhile controlling something that you feel you can't afford to have.)

5. An inventory level should be sufficiently maintained to service 95 percent of the orders from stock upon receipt, with shipment made in three days. The first point that must be recognized is that a company can't possibly afford the inventory it would take to have everything on hand for every order it would get. The second point to be considered is providing reasonable customer service by allowing sufficient time for filling the order.

It must be recognized that any one of the five criteria above might be met without a great deal of effort. But having just one without the others could cause a company to miss by far whatever profit objectives were set. So inventory control requires the proper blend of procedures to meet a number of balanced goals.

The standards listed above are not to be considered irrevocable. Certain points may need to be relaxed. It would be a good idea for management to publish its objectives so that all concerned will know their goals.

When this point has been reached, the analyst will know what management expects from the system. He will know against what he will be judged and how to determine what all operating people need to produce those results.

Role of Organization Chart

Upon seeing the published list of inventory objectives and recognizing that the analyst is undertaking a program to produce such results, the specific people concerned with inventory control are not necessarily going to phone him right away to let him know their exact requirements or volunteer to provide all the help they can. There is always going to be a certain amount of inertia among people. (In a sense, the analyst may be looked upon as an enemy because he is setting out to do something others didn't accomplish.)

If you are a systems analyst, you may have to search out the necessary people, which means you must obtain information about duties throughout the organization. Care must be taken not to frighten people about the complete reshuffling of duties. You must learn enough about an area so you can properly study it in relation to all other areas and get into a position to recommend the types of things that should be done. It could be that all the raw data that you need to produce the results above is available right now, and perhaps you just need to rearrange a few steps and add certain analyses here and there.

The organization chart of a typical manufacturing company appears in Figure 2-1. The name of each functional area is very suggestive of the general duties of that group. Organization charts often contain the supervisor's name in each block; this information can be very helpful, since it is desirable to learn the name of the supervisor of every person you work with. You will find it is a good idea to speak with a person's supervisor before seeing the worker, so that the supervisor is informed of your activities. It is also advisable to obtain job descriptions of various people, if that is available. In fact, it is often the responsibility of the Systems Department to prepare and maintain job description manuals.

Not all companies prepare organization charts and some of those that do will not make them available for purposes such as this. In such cases, you must figure out other ways to obtain the needed information. One way you might try is this. Once you think you have located the specific person with whom you want to deal, you may say, "Please let me have your supervisor's name so I can clear this with him before we get started." Or you may tell the employee that the manager of the Systems Department would like to briefly describe why it is necessary for you to obtain certain information from them. Whichever way it is done, it is important that people receive prior knowledge of your activities.

Questions

1. Approximately what rate of return would you require if you put your money in a bank savings account? U.S. Government Bonds? AT&T stock? A business of your own? Assuming that you quoted varying rates for the four different situations above, why did you do so?

2. Why might a person be willing to set up and operate his own business for only $12,000 profit a year when he could work for someone else at $18,000 annually?

3. One of the foremost business objectives should be to increase sales. True or false? Why?

Figure 2-1.   Organization chart.

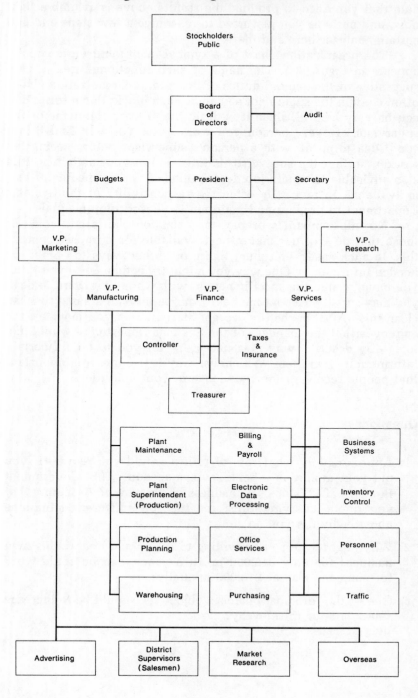

4. What are the basic reasons why a railroad freight car spends so little of its life in motion. What do you suppose it would be worth to a railroad if you could show them a solid way to double that figure? What have airlines done to keep planes in the air as much as possible?

5. Why might data processing specialists be allowed to determine what information will be processed? Who should be the one to determine it?

6. Prepare a very broad list of objectives for a payroll operation. Then for each item listed, give a very specific quantitative value.

7. This chapter listed five very specific quantitative objectives in an inventory control situation. One of them was to ship at least 95 percent of all orders within three days of receipt. Aside from what might be done in a sound way to reach that objective, indicate an "unsound" thing that might be done to reach that objective. Then clearly indicate the adverse effects it may have on each of the other four objectives.

8. Why might a company hesitate to prepare an organization chart, even for internal use?

9. Auditing and budgeting are commonly considered to be financial activities. Why might a company want to have those functions report to the president or to the board of directors rather than to the vice president of finance? (See Figure 2-1.)

10. If you were the treasurer of a company, what objectives might you have concerning the general cash level? For instance, to what other accounts, if any, would you want to relate cash? Assuming it were possible to accumulate that much, would there ever be a point at which a company might have too much cash for its own good?

11. One of the goals of the U.S. Post Office is to provide next-day service to all points in the country on first class mail. If this should happen, how much of the credit should go to ZIP code?

12. Suppose you had heard that ABC Grocery Chain makes only $1^1/_2$¢ on the dollar. Obviously they could earn more by just putting their money in the bank. True or false and why?

13. Can inventory turnover be calculated for a company on an overall basis by dividing the total number of items sold by the

average number of items in inventory? Why? Can inventory turnover be calculated for an individual item by dividing the dollar value of what was sold by the average dollar value in inventory? Why?

14. A billing and accounts receivable system generally involves the steps of invoicing customers for charge sales made to them, collecting their payments, and keeping records of how much each customer owes. Recalling the five fairly precise objectives that top management set for inventory control in this chapter, establish precise objectives for each of the following points in billing and accounts receivable.

   a. What is the maximum number of days after a sale has been made that the invoice must be prepared and mailed?

   b. What is the maximum number of days after receipt of a customer's payment that the check must be deposited in our bank account?

   c. What is the minimum acceptable accounts receivable turnover? (That figure can be calculated by dividing charge sales by average accounts receivable.)

   d. What is the maximum percentage of total assets that accounts receivable is allowed to be? Justify why you picked the value you did.

   e. What is the oldest you will allow any unpaid bill to become before you turn it over to your Legal Department?

Other Readings

Corporate Financing, March-April, 1970, p. 43. "MIS: What Has Gone Wrong?" Discusses what has happened to the "M" in MIS (Management Information Systems), pointing out that not enough systems are designed on the basis of what is needed.

Newman, William H., and James P. Logan, *Business Policies and Central Management*, 5th ed. Cincinnati, Ohio: South-Western Publishing Co., 1965. Chapter 4 reviews basic company objectives.

CHAPTER 3

Identification Codes

Identification codes are used so commonly that we may tend to overlook their existence or fail to realize the impact they carry. Names of cars, street addresses, and even our own names are examples of codes. Even though much of the systems work you might do would require that you continue to use an existing coding structure, it is important that you have background in code development so that you can apply the idea at the proper time.

In a practical situation, you might not become concerned with the coding structure until you had reached the systems design stage, but it is introduced here so you are able to use it all through the study. I believe the topic is of enough importance to devote a whole chapter to it. If you have a checking account, savings account, and mortgage at the same bank, it is likely that each has its own identifying number. For several reasons it is desirable that the same number be used for all activity with each customer.

In business, we must have highly effective ways of identifying the people, companies, places, and things we deal with. Airlines find it convenient to refer to flight number "421" rather than saying, "The Lufthansa plane at 10:15 P.M. from Boston to Frankfurt." A car rental agency may use car number "475Y" to refer to the "blue 1971 Chrysler four-door hardtop." The U.S. Post Office can send a letter to ZIP Code "02154" more easily than if it refers to "Waltham, Massachusetts." The code does not replace or eliminate the full description; the code is merely a shorthand method of distinguishing one flight or car or post office from another. The code is especially helpful when you are about to file data, retrieve it later from storage, communicate regarding it, bring like items together, or when you are about to put all items in sequence by sorting.

Thus, a "two-inch piece of grooved metal with a tapered shape and one wide flat end used for holding wooden things together" may be called a "2-inch screw, #12748." The code is generally set up to be shorter than the words that would otherwise be needed to describe the item. The code might be more meaningful and it must be set up in such a way as to make it easier to pull like items together into desirable categories.

Alphabetic Versus Numeric Codes

An alphabetic code has the obvious advantage since it can come closer to the manner in which we most often name things. To this extent, a "rope" is more meaningful to our way of thinking than part number 17629. But to refer to a "400 hp alternating current motor" may be too inefficient a use of the means available for recording. When a shorter code is needed, the tendency has been to develop an all numeric code rather than to develop an alphabetic abbreviation.

Much of the tendency to use numeric codes has occurred because early technicians were not able to design machines that could handle alphabetic characters. It was just too difficult to develop the necessary circuitry. In 1961 I phoned the local hotel in a nationwide chain to have them make a reservation for me in another city. Since their method of sending data to the destination point was not carried out by a voice telephone call but by a wire transmission method, they could not send my name to the hotel; instead they used a fairly random code—the last four digits of my home phone number. When I eventually arrived at the hotel, I said I was "3719," and they assigned a room to me. It is certain there would occasionally be duplicates of these codes, but the method apparently worked, for the hotel had used it for many years.

Even when it became technically possible to handle alphabetic characters, the use of them often causes operating problems. For instance, it takes twice as many passes to sort alphabetic characters on a punched card sorter as it does numeric characters; it takes somewhat longer to keypunch and there is basically more chance of error with letters; and verification processes such as hash totals and self-checking digits are much more complicated. (Hash totals and self-checking digits represent special techniques for verifying the accuracy of data. Most data processing texts give illustrations.)

Practically speaking, most computers now using magnetic tapes and disks handle alphabetic and numeric characters equally well. But since most of them still use punched cards extensively for input, the numeric coding has remained the prominent method. Many users are also reluctant to change extensively from numeric to alphabetic characters because of the great cost and disruption that would occur in their organizations. It is not as great a change as would be involved in switching to the metric system of measurement, but the details would be significant, and, many feel, not worth the value to be gained.

Our everyday lives have become proliferated with numeric codes for most people. For instance, we have a five-digit ZIP code, a ten-digit phone number, a nine-digit Social Security number, an eight-digit checking account number, and a ten-digit oil company credit card number. Any magazine we subscribe to may keep track of us by a mixture of alphabetic and numeric (alphanumeric) characters of perhaps staggering length. For example, the code on one of my subscriptions contains 30 characters over and above those for the name and address.

The whole topic of numerically coding transactions and customers has caused concern among many people that we as individuals are being de-humanized and turned into numbers. One group of people in California protested this to such an extent that they went to the Supreme Court to prevent the telephone company from converting any more of the AAN-NNNN(TY1-2000) type of phone numbers to the all-numeric (891-2000) style. After a one-year wait, the Supreme Court could see no further reason to prevent the change, so they reversed their earlier decision.

Considering the typical person's relative lack of knowledge with respect to data processing matters, it is little wonder that citizens can be easily stirred by the de-humanizing aspect. But people must be educated on this point. It doesn't take much time or effort to adequately describe why a code is needed. The following statements can serve as sound justifications:) SAVES TIME, MONEY

1. The telephone company decided to switch to all numeric phone numbers as the number of customers grew rapidly and as certain "numbers" clearly became not acceptable. For instance, in an area where the first two positions of the number were composed of the second and ninth codes on the phone dial, some customer who makes saws may complain if the code were "AX" and someone else may complain if the code were "CY" or "BY." But "29" is not so objectionable, and the approach has provided for more usable numbers. However, it does not provide for more numbers, as there are only so many combinations on the dial.

2. It is technically feasible for the telephone company to set up codes in which you could call the destination point by dialing the person's name and a portion of his address. But the extensive requirement for the necessary switching facilities would greatly increase the cost of phone services, and it would take considerably longer for you to dial each call.

3. For most of the 90,000,000 federal income tax returns filed each year, the regional IRS Center can key stroke each nine-digit

Social Security number to identify each return. Try to imagine the increased time required if the person's name were used, since his address or something else would also have to appear to differentiate among people who had the same names.

⌐ 4. Notice that most any credit card you may have contains an account number and your name in raised form on the card. When a charge slip is prepared and eventually processed, it is optically scanned by a machine for input to a computer. It requires a less costly machine to read the account number than it would to read your name and address.

The use of a numeric code by any group has not in itself reduced the importance of a person's name. For instance, because of a ZIP code, has there been a trend toward leaving an addressee's name off an envelope? Has the bank teller quit thanking you by name for your deposit just because your account now has a number? Has the Internal Revenue Service deleted space for your name on your income tax return now that your Social Security number is the major identifier to their computers? Do you have any credit cards that contain only your account number and not your name? The code is merely a less expensive way of providing you service.

Meaningful Codes

A basic decision that has to be made on codes is whether or not certain portions of the code are to have special meaning for any purpose. A typical way to develop a code with no significance is to assign codes consecutively; that is, the first person hired is clock number 00001, the second 00002, and so on. This approach is especially useful where you might want to publish a list of employee's names in seniority order.

But there are several reasons why you might want to assign a special prefix to categories of people. In order to collect payroll costs by specific areas of the business and in order to hand out payroll checks in an efficient manner, it might be desirable to issue department numbers as well. Thus, if the Accounting Department has been given number 050 then man-number 050-00001 is the first person hired in that department. In payroll then, you may have two different coding systems. Social Security number will satisfy U.S. Government requirements, and clock number will help provide important internal information.

(As a specific example of a meaningful code, General Motors Corporation adopted a 13-position serial number for vehicle identification several years ago.) The positions of the code have meaning as follows.

1st, 2nd	vehicle series
3rd	base engine equipment
4th, 5th	body style
6th	model year (units position only)
7th	assembly plant
8th–13th	sequence number of car built that year

A car with a serial number of 164391 F 000541 would represent as follows.

16	Chevrolet
4	V-8 Engine
39	Impala 4-door Sport Sedan
1	Built in 1971
F	Assembled in Flint, Michigan
000541	541st 1971 Chevrolet built in Flint

One company in the wholesale business uses an alphabetic code very effectively. It is their wish that their own sales personnel have ready access to their internal product cost. The company has chosen the ten letters of the word CHARLESTON to represent the digits 1, 2, 3, . . ., 9, 0, respectively. Thus, the following entry may appear in certain sales literature:

"Widget, black 4" by 2½" (RET) 8.75   11.50"   ⟨ DIRTY-WORD CODE

This means that the wholesaler's cost is $4.68 (R = 4, E = 6, T = 8); the price to the retailer is $8.75; and the suggested retail price is $11.50. Thus, the salesman can show the advertising to his customers without the fear of their learning how much gross profit the wholesaler is making.

(Basic misunderstanding by customer departments of what is involved in the use of codes has plagued many data processing departments. Many payroll systems stick so faithfully to the meaningful-code approach mentioned above that they issue the number of a person who has left the company to his replacement. You can see the obvious disadvantages of a practice like this, as you can never get true meaning from the pay collected against that clock number.) This is another area where management participation will help iron out problems and provide the boost to arrive at acceptable solutions.

Another problem can occur with respect to the person who may be transferred frequently from department to department. In one company this happened often as severe budgeting problems developed. Certainly the Data Processing Department shouldn't have the power to say the practice couldn't be allowed just because it creates a little more work for them; on the other hand, the practice may provide advantages and great flexibility for operating departments. The two groups must work together to provide flexibility to operating departments and also to assure that procedures are reasonably simple for data processing to keep track of the people.

Need to Develop Codes

It is conceivable that you could spend many years as a systems analyst and never have the opportunity to develop any codes. This could happen because a company has no doubt been using codes on all its applications for many years, and there may be a great reluctance to change. You aren't likely to be too popular if you suggest that your company adopt a whole new part number or clock number structure for what would appear to provide few benefits.

(Significant reasons for wanting to change codes might include the following.)

1. A new type of mechanization might require different codes. For instance, using 01, 02, etc., for months would make it convenient to sort data into monthly order whereas an alphabetic sort on month would be meaningless.

2. More significant information may be needed. You may want to add a product group prefix to all items so they can be pulled into meaningful groups.

3. Merger with another company and blending the structures would require new codes.

4. Installing a common system within the divisions of a corporation might become necessary.

(If you should arrive at a point where a drastic change seems desirable, you must carefully assess such benefits in relation to the disruption that would be caused by a change.) Perhaps you could informally determine from several people you can confide in what the implications of a new coding system would be. If the conversion can't be realistically made by working through everyone, then perhaps the desired benefits can be obtained by working around certain people. Adding to the existing code within the data processing area may be done in such a way that all the operating departments can receive some benefits but not be bothered with all the details. For instance, suppose someone needs statistics of sales and profits by county, but county code is not included in the present customer number. Instead of changing the existing code, perhaps you can develop a cross-reference method that will relate county code to customer number. Assume that all sales are identified by customer number. The Data Processing Department might maintain its cross-reference file like this.

Customer No.	County	Customer No.	County
12497	019	13085	142
12499	321	13097	321
13076	001	14513	001
13081	019	16940	142

Identifying the sales by county is a matter of matching county codes to customer number codes. The sales transactions can then be sorted to county code for the purpose of preparing the required reports.

The time spent in processing the cross-reference steps over the next five years may be considerably less than the time spent arguing about a complete change. (Again, if the benefits of a change look great, top management should assume a role in seeing that they are realized.)

If a code is to be completely revised, then the timing of the change should be carefully considered along with all other activities. For example, if it is being done in conjunction with obtaining a computer, then try to do the proper educational and trial runs to iron out code change problems early enough so they don't exist at the same time as computer conversion.

Inventory Control

Since the purpose of the inventory control system is to collect and analyze information about the items you are buying and selling, it is therefore necessary to develop a way to properly distinguish among items. In how many ways might you wish to identify an item? You might do it by:

1. an alphabetic name—a screw, a bolt, etc.
2. a unique number—#147 or #198436.
3. its introduction date—1958 or 2/7/64.
4. its source—purchased, manufactured, or both.
5. potential profitability—cost related to selling price.
6. how it is packed—by the dozen or gross.
7. location—warehouse 4 or bin 2.
8. major use of product group—fastener or tools.
9. intended customers—retail, consumer, or manufacturer.
10. degree of usage—high, medium, or low.
11. seasonality—winter or summer.
12. person responsible for its control—analyst 5 or analyst 6.

You can see how the overall identifier would get longer and longer as you add more sections to provide meaning. But recognize you may record a transaction against an item by indicating only the unique number (number 2 above). Then when the record for that item is retrieved from the file, all the other data is there.

Each portion of the code must be analyzed to determine if the cost of providing it is justified by its value to the organization. If the company deals in only six items that are truly of a seasonal nature, perhaps they can't afford a special code for seasonality. In such a case it doesn't seem to be asking too much for people to remember which the seasonal ones are. If a company is dealing with 20,000 items and has thirty inventory analysts involved in their control, a two-digit code identifying the responsible analyst may be completely justified.

Note that if the twelve individual portions of the code are to have real significance, each portion must be brought up-to-date as conditions change. It may have seemed like a good idea to show location as part of the code for each item, but if warehousing procedures can move an item about constantly with reasonable control, then perhaps too much time is being spent changing the bin number in your records.

A code may also be used as a good shorthand way of recording error conditions. Suppose the quality control section of an assembly operation wishes to keep track of the incidence of certain defects. Thus, it is relatively simple for the inspector to write "01" when the paint job is poor, "02" when a bracket is missing, and so on. Here are some of the advantages of using a code rather than longhand.

—1. Recording the defect would be much quicker.
—2. Precise categories of defects can be developed and statistics gathered against each style in order to pin down problems and responsibilities.

These are some of the disadvantages of a code.

—1. The inspector will have to take time to look up the code.
—2. He may make recording mistakes when relying upon his memory to determine a code. A mistake would not be so likely if he just wrote "bad paint."

In manual systems, the forms that have been used to record inventory transactions can have descriptive headings or they can have other distinguishing features such as color and format. These visual differences alert people to the types of transactions they are and help to process them accurately. An automated system will require each transaction to have a separate identifying code, not only to distinguish among items but also types of transactions. This is not so much for the pure purpose of inventory control (that of knowing how much you have), but for the other purposes which can be served by the workings of inventory control. For instance, you need a separate code for:

—1. receipts from your source of supply (whether made yourself or purchased).
—2. sales.
—3. returns by customers.
—4. debit adjustments. This may come about by taking a physical inventory and finding there are more on the shelf than what you can account for from your records.
— 5. credit adjustments. The reason might be the opposite of that in (4) above.

Notice that transactions 1, 3, and 4 all have the effect of adding to the balance on hand. But separate statistics may be needed on each category of transactions, and this is why code (1) can't be used for all transactions that increase inventory. Perhaps management wants to know what total returns are in order to determine whether customers are being oversold, or a separate accounting for returns may give an indication that products are being returned because of quality problems.

If sales transactions are to be broken down further into lease sales, sales on an installment plan, and issues on consignment, separate codes will be required for each.

Questions

1. Why would it be desirable for a bank to use the same identification code for a person's checking account, savings account, mortgage, and personal loans? Why didn't all banks set up these codes in this way from the start? Why haven't they converted to one code per customer by now?

2. Why are codes needed?

3. Why hasn't there been a big switch to alphabetic codes now that equipment can handle them much better than it could ten to fifteen years ago?

4. How many digits are there in the social security number? Is there any significance to any of the positions in the code? How many people will the code identify, assuming a number is never given to another person? About how many of the numbers have already been used? Project roughly for how many more years there will be enough codes available under the present scheme. How do you suppose the Social Security Administration will expand the code when all the present numbers have been assigned?

5. Carefully explain why the telephone companies are converting phone numbers to seven digits (eliminating the two alphabetic exchange letters in the first two positions).

6. Why might a company choose to set up its own clock number as an additional code when it is legally obligated to keep track of employees by social security number?

7. What basic pattern exists in the way that U.S. Highway and Interstate Routes are numbered?

8. Name two very significant classifications which might be desirable in the credit card account number of a department store which has customers and offices nationwide. Explain.

9. In what situation would it be very desirable to assign just plain consecutive clock numbers to employees? (The company still uses social security as required.)

10. Write a brief letter that might be sent to customers telling them why you are setting up a customer number as opposed to identification by name and address only.

Other Readings

Coding Methods, Form F20-8093, International Business Machines Corporation, 112 East Post Road, White Plains, New York 10601.

Withington, Frederic G., "Cosmetic Programming," *Datamation*, March 1970, p. 91. This article presents guidelines as to how computer output could be printed with meaningful explanations and terms rather than just numeric codes which the reader would have to memorize or look up.

CHAPTER 4

Forms and Report Analysis

In the logical process of developing a system, it is better to forego forms and report analysis until at least the present system has been studied. You would then be at the point of determining what is needed by all the people who have responsibility for operating the business. But for the purpose of this book, the material has been presented earlier in the hope that you will be better prepared to study the present system. For the first few systems you work on, it will be helpful if you know as much as you can about the interrelationships of various segments.

No matter how sophisticated a system may be in terms of computer hardware and telecommunication devices, there are countless places where reports will be required in the processing cycle. It is true that a lot of paperwork is used as a defense mechanism for the operating people, but if you aren't successful in designing systems that will eliminate much of that paper, you are obligated to help them produce and work with paper in the most effective manner.

The following are some representative examples where generation of paper to back up a transaction is needed.

1. A receipt should be given to a customer whenever he makes a cash payment.
2. A vendor will require a purchase order which is a formally prepared document containing an authorized signature.
3. Good business practice calls for a packing slip to accompany goods shipped so the recipient can immediately identify what he received and also be able to report any deficiencies.
4. An employee expects some details about the calculation of his gross and deductions to accompany his paycheck.
5. Paperwork often represents a form of approval, such as the ticket to get on a plane or a prescription to take to the drugstore.

After preparing a form to give someone else, it seems only natural that the preparer keep a copy or so for himself. In this way he can know at all times what has previously occurred regarding that transaction.

Although some systems have been highly successful in cutting down on paperwork, the rapid rise in use of paper and the staff to process it are evidence that this is an ever-growing segment of

51

business. Everyone seems to need more information, and information implies paperwork. In 1958 the paper industry predicted that paper consumption would reach 47,000,000 tons by 1975, but that usage was actually reached by 1968.

A common statement in business operations is to "put it in writing." And since many people wait to perform their work, or wait for some corresponding physical act to occur, or need to keep a record of what has happened, they need paper to show what has to be done or what has already been done.

In a 1966 report entitled *How to Cut Paperwork*, a committee of the U.S. House of Representatives disclosed the following facts about federal government paperwork.

1. There were 360,000 different forms.
2. The cost of printing the forms in 1966 was $53 million.
3. About one-fourth of all records were to be kept permanently.
4. The federal government had 255,000 employees whose main job was filing records.

Forms and Report Costs

The typical approach to discussing forms costs is to talk about the printing cost of a particular form or to summarize the purchase costs of all forms. This is a narrow-minded approach because it tends to overlook the high costs of preparing and processing all the forms that are used and any "costs" of not having the proper forms. Published studies point out that, on the average, the cost to buy a form is only about 5 percent of its total effective cost. So if a company spends $50,000 a year to purchase forms, there is great reason to believe they are spending about $1 million a year processing them. If nothing else, this should shock many people into realizing what the processing of paper really costs them. Processing includes such steps as buying (the steps involved, which includes paperwork), receiving, storing, filling in, mailing, stapling, sorting, filing, retrieving, looking at, and eventually throwing away.

Another important point to consider is how purchasing costs rise as you obtain more copies of a multi-copy set. The following price quotes are typical for a lined computer output form, 11 by 15 inches:

Parts	Price per 1000, Lots of 10,000 Sheets
1	$ 7.50
2	20.10
3	32.80
4	44.50
5	58.00
6	75.40

Great care must be taken in determining how many parts each form should have. On the one hand, you can see how costs of multi-copy sets increase. On the other hand, if six or eight copies must be realistically prepared, there is a great chance that there will be little legibility beyond the third or fourth carbon, and several processing runs might be required to get sufficient legible copies. One large company discovered that 95 percent of all reports had use on a one-time basis only. If the typical company has a similar experience, perhaps systems design can provide for more sharing of reports than running off a copy for each potential user.

Forms Control

At this early stage in the development of a system you should not be spending too much time on the details of gathering data on transactions. But it is desirable to at least be thinking of the problem of forms control, both with regard to the source documents employed and reports generated by your system. This control involves procedures necessary to assure the following.

1. Realistic means must be brought about to prevent the creation of more forms without a sound analysis of their value. For this reason forms design and control must realistically follow behind systems design; if it does not, you are likely to design the system to utilize all the forms.

2. Forms must be designed to provide what they are supposed to and have nothing inherent in them which tends to create errors or misunderstandings.

3. A central issue point or clearing house is necessary. Some companies assign a person the duty of forms controller. Perhaps all he does is issue a consecutive number to each form that a user designs and then retains a copy of each form in his own file. My

concept of forms control is one where procedures are carefully reviewed to see if forms are necessary. If they are, the controller tries to see if there may be an existing form which can do the job or perhaps be altered to do the job. (Forms control should be a coordinated effort, just as is systems design. The controller should try to prevent the widespread use of the "bootleg" form, one that a user makes up on his own without relating to what else may be available or the potential needs of others.)

   * 4. Proprietary or government classified documents must be safely stored.

   * 5. A realistic disposal program is planned.

   In business it can become so easy to reach a point where the solution to an alleged problem is saying, "What we need is a form" and then to design one. This method of operation, if allowed to continue without proper control, will result in every department going in its own direction on forms design. A form, or a very similar one, will probably be invented many times and the various departments will never be tied together in a compact system (If a central issue point is established within the Systems Department, great duplication should not take place, and the needed form can best be designed to serve all users.) This is also the most effective place to prevent the birth of a new form if it can not be justified by necessary procedures.

   A student once came to me and said he had received a phone call at home from the treasurer's office of his employer stating that he was going to be sent to jail. He was told he had cashed his pay stub. He told the caller that was ridiculous because he had the stub in his wallet. He reached for his wallet to pull out the stub but was surprised to find he really had retained the check instead. Whoever had designed the check and stub made them look alike. In the rush of a busy day at the bank, both the worker and the teller failed to notice they had processed the stub.

   In 1968 a number of banks began offering their checking account customers "scenic" checks displaying orange groves, mountain ranges, and the like. In many cases, the scenes were so large they partly covered space normally reserved for amount and signature; the ink used was so bright it reduced the ability of microfilm equipment to record each check processed. Some operating problems became so great that many banks were forced to withdraw the checks or to considerably alter their design so the checks once again became of practical use for all phases of the operation.

Both of these illustrations clearly show that forms design is not something that can be done in a casual manner. It requires very detailed work by someone who is familiar with the various processes that are going to be followed in preparation and use of the form.

(Many business documents are inherently valuable and must be carefully guarded against falling into the wrong hands. Certain purchased, but not filled-in forms, such as payroll checks and accounts payable checks, must be closely controlled to prevent use for other than their intended purposes.) This is often accomplished by having the check printer prenumber the checks and then locking them in a place under strict control of one authorized person. The person about to use a quantity of checks may be required to sign for them, with the custodian taking care to see that control is maintained between the starting number of those issued on a particular day and the ending number issued on that day. A careful count must be made of the number of checks used along with an account of any voided ones. Check reconciliation procedures should later help to uncover any discrepancies, but reasonable steps must be taken to prevent problems as well as catch them.

Various reports generated within a company contain information which management does not want made available to other companies. Included are such things as detailed cost figures, individual's salaries, sales projections, and profits on individual items or at division levels. These very things are also usually withheld from certain employees. This is accomplished by using carefully planned and controlled distribution lists, locked files, special messengers, and so on. Another way is using a form designed in a similar fashion to the one in Figure 4-1. Through the careful use of the black dot pattern on certain carbon copies, information that is not to be disseminated to certain employees will not be legible; but use of such a form means the report has to be prepared only once.

If the information is to be kept truly secret, then procedures must be set up, understood, and followed. One supervisor would not let his secretary type his recommendations for bonuses; he typed them himself. But due to his own habit of never using a carbon paper more than once, it was like broadcasting to the world when he threw the carbon away.

(A rubber stamp marked COMPANY PROPRIETARY should be used to alert everyone of the expected treatment of a document.)

Figure 4-1.   Form design prevents print-through of dollar information. (Courtesy UARCO Incorporated, Burlington, Mass.)

CUST. ORDER NO.	SALESMAN	SHIP VIA		TERMS	DATE	
QUANTITY ORDERED	PART NUMBER	DESCRIPTION		QUANTITY SHIPPED	PRICE	AMOUNT

FORM ES 555-3 INV   THE MERCHANDISE LISTED ABOVE HAS BEEN PRODUCED IN ACCORDANCE WITH THE FAIR LABOR STANDARDS ACT OF 1938, AS AMENDED.   **ORIGINAL**   ▲ PLEASE PAY LAST AMOUNT SHOWN

CUST. ORDER NO.	SALESMAN	SHIP VIA		TERMS	DATE	
QUANTITY ORDERED	PART NUMBER	DESCRIPTION		QUANTITY SHIPPED		

FORM ES 555-3 BO   THE MERCHANDISE LISTED ABOVE HAS BEEN PRODUCED IN ACCORDANCE WITH THE FAIR LABOR STANDARDS ACT OF 1938, AS AMENDED.   **SHIPPING COPY**   ▲ PLEASE PAY LAST AMOUNT SHOWN

▲ THE FORM SHOWN ABOVE IS A UARCO CUSTOMER PROVED DESIGN ▲

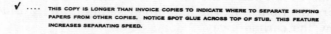

ONE WRITING TO PREPARE THREE FORMS . . . .

✓ . . . . . . GET WORK UNDER WAY FAST - WRITE ORDER, INVOICE AND SHIPPING PAPERS AT ONE TIME. DETACH SHIPPING COPIES. OTHER COPIES CAN BE HELD IN FILE WITH CARBONS INTACT SO THAT SHIPPING INFORMATION, SHIPPED QUANTITIES, PRICES AND EXTENSIONS CAN BE ADDED LATER.

✓ . . . . THIS COPY IS LONGER THAN INVOICE COPIES TO INDICATE WHERE TO SEPARATE SHIPPING PAPERS FROM OTHER COPIES. NOTICE SPOT GLUE ACROSS TOP OF STUB. THIS FEATURE INCREASES SEPARATING SPEED.

A paper shredder should be considered as a possible means of making sure that discarded forms are not still legible.

A company can spend a great deal of money retaining forms long after they have served their useful purposes. Of course, there are certain legal requirements that must be met. But there is a great deal of hearsay on this point, and the average person usually thinks the legal requirement is much greater than it really is. A forms retention and disposal program should be set in motion as soon as a form is designed; because of this you may be able to eliminate keeping the form at all after it has served its initial purpose. The entire program must be viewed from the risk standpoint; what risk are you subject to by not having it as opposed to the cost of keeping it?

Following are examples of what some users have been able to do in order to eliminate paperwork or some processing directly related to it.

1. Quarterly Social Security reports and annual employee earnings statements can be sent directly to the U.S. Government on magnetic tape rather than in printed form on paper.

2. Many companies deposit a person's salary directly into his bank account. The employee receives the normal earnings statement; but the company has to prepare only one check for deposit in each bank involved. Check reconciliation and storage are virtually eliminated.

3. Voice output is being used in many systems. The New York and American Stock Exchanges both installed voice output years ago to enable interested people to obtain current prices from a computer that answers from a recorded voice. No special output devices were required because every user always had a telephone. On some long distance calls placed from phone booths a pre-coded voice tells the caller how much money to deposit; this has eliminated the need for a telephone company employee to look up the charge in published tariffs.

4. Output from cathode ray tubes (CRT's), which are similar to a television screen, is now very common. The user obtains what information he needs when he wants it without generating paper.

5. The banking industry is taking a close look at what has been called the "checkless society." The term does not mean no checks at all; it really means fewer checks. When you would buy something at a store, you would present an identification card to the merchant. He would insert it into an input device, key in the

amount of the sale, and then communicate over a telephone line to the bank computer. It would deduct that amount of money from your account and put it into the merchant's account. You could still get a receipt from the merchant and a statement of your account from the bank, but the processing of that costly form, a bank check, would be eliminated.

6. A very clever forms design is shown in Figure 4-2. It is a long, continuous form composed of four major parts.

Figure 4-2. Sample of a computerized form that needs no stuffing prior to mailing. (Courtesy UARCO Incorporated, Burlington, Mass.)

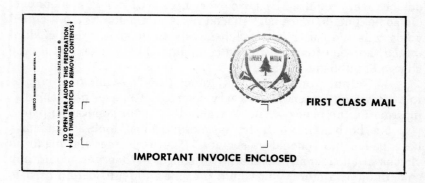

(a) Part A is the original, in this case a bill to a customer, with a section just beneath the premium charge for name and address (shown at middle of page 58).

(b) Part B is a small section of carbon paper (not shown).

(c) Part C is an actual envelope (shown at bottom of page 58).

(d) The envelope in part C contains the actual invoice form that will be mailed to the customer. (The invoice form was pre-stuffed by the manufacturer of the form.) The inner side of the envelope is made of carbon.

The form is manufactured and assembled with B placed between A and C so that the name and address portions of A and C line up with the carbon portion of B. A computer prints the billing information and the name and address on part A. The carbon on part B permits the name and address only to print through on the outside of part C (the envelope). The carbon inside the envelope is large enough to allow all information being printed on part A to also print on the form inside the envelope.

After the printing pass through the computer the forms are decollated and the carbon (B) discarded. Form A is retained by the company as their copy of what was printed; part C is burst into individual envelopes and mailed.

Inventory Control

Let's suppose that after we have been successful in getting the inventory control objectives from management, we do a certain amount of studying to understand what inventory control is all about. This is not for the purpose of deciding what data will be processed, nor is it to arrive at any conclusions as to what the new system will be. It is for the purpose of knowing the subject well enough so that we can do the best job possible. Since much of the process is going to involve asking questions and interviewing people, we must do what is necessary to be prepared to do a good job of that. We must assume the operating people do not know everything about their jobs, and we should get in a better position to be able to make worthwhile suggestions.

From what you already know about inventory control, perhaps from reading several articles and books and from informal discussion with people in the organization, you might develop a list of tentative departmental requirements as shown below.

Department	Requirement
Advertising	Excessive quantities of items on hand so advertising campaigns can be properly prepared. New items that will be available for marketing soon.
District Supervisors	General profitability of items. Average time needed to ship an item when not in stock. Is item in stock now? Back ordering problems.
Market Research	What new products are being added? Will they replace existing items?
Overseas	Similar to District Supervisor's above.
Plant Maintenance	Availability and location of maintenance items.
Plant Superintendent	Information on what and how many to make and when. Availability of raw materials. Are new manufacturing processes needed?
Production Planning	What is to be made and in what quantities? Emergency orders.
Warehousing	Size and bulkiness of finished items to figure storage space requirements. What are the fast movers so placement can be nearest the front door?
Controller	Dollar values for statement purposes. On hand, on order, and sold quantities.
Taxes and Insurance	Amount and location for personal property tax and insurance purposes. Any unusual shipments requiring unusual insurance protection.
Treasurer	Purchasing and payment due date requirements to make proper funds available.
Billing and Payroll	When, what, and to whom was a shipment made? What freight and tax situations apply?
Electronic Data Processing	Everything to be processed in a mechanized form must be made available to this area.
Inventory Control	Costs of providing vs. storing. Desired levels of each item. Location. How much to buy?
Purchasing	What, when, and how much to buy? They must maintain their own data on vendors and prices. What alternatives are there to obtain needed items? Does the company have the means to purchase enough to obtain quantity discounts?
Traffic	Size and weight of items and how they are packed. Urgency required.

Questions

1. Have computers done much to reduce the need for reports and records? Why?

2. What is meant by "forms control"?

3. What is a "bootleg" form?

4. Why would a bank offer its customers scenic checks? What problems might develop from their use?

5. What is meant by "proprietary" information? Who is responsible for its safe-keeping? What is meant by "Classified" information?

6. What is so desirable about voice output?

7. What is the "checkless society?" Will it eliminate checks?

8. Has most of the increase in paperwork been due to the increase in the population? Why?

9. Shown below are two examples of printed reports. Style A is known as a detail printed or listed report. Notice there is a line for every item and a line for every total. Style B is a group printed report. For the same input data only totals are printed. Give the advantages and disadvantages of each.

Style A			Style B		
Dept.	Clock No.	Amt.	Dept.	Clock No.	Amt.
12	1	5	12	1	11
	1	6		2	12
		11*			23*
	2	4	13	8	7
	2	8			
		12*			
		23**			
13	8	5			
	8	2			
		7*			

10. A systems analyst designed a section of a form as shown below to gather pertinent data. Make a basic redesign so that recording the data is much simpler.

College grad.?	Yes ☐ No ☐	
If yes above, year	_ _ _ _	
Married?	Yes ☐ No ☐	
If yes, spouse's name	_____	

Other Readings

Marien, Ray, *Marien on Forms Control*. Englewood Cliffs, N.J.: Prentice-Hall, Inc., 1962. This excellent book covers the entire field of forms control.

Olsen, John L., Jr., "Forms Design and Control," *Ideas for Management*, 1962. This article concentrates on the principles of good forms design and presents a valuable approach to help determine if a form will pay its own way.

CHAPTER 5

Files and Contents

A file is merely any one of many possible types of containers in which certain records are going to be stored for a period of time. There are two major categories of files with respect to their eventual use.

1. Files that contain the results of past transactions are being kept because of the need to refer back to them. A company may need reference to sales made to customers over the past few years, so they save a copy of the invoice relating to each sale. Also, laws require that certain payroll records be retained for three years. Since a transaction can not necessarily be forgotten and evidence of it destroyed as soon as it has occurred, businesses have had to establish and maintain files.

2. Files that contain those units of data that will help to process future transactions such as next week's payroll; it requires the retention of everyone's pay rate, year-to-date earnings and taxes as of the last pay, as well as information regarding each person's deductions. Instead of disseminating all this information to the points where future transactions will occur, such as to a time clock for the next recording of a person's time, the transactions are processed with as little detail as possible and then moved to the file. In payroll, all you may need to record is a worker's identification code and the hours he worked. Hopefully this practice will keep transcribing and moving of data to a minimum and also assure that the file information is always secure and readily available for processing.

The major considerations involved with files are:

1. whether there should really be a file, and if so, what should be in it.
2. the form in which data will be stored.
3. what is the method of accessing and processing transactions against the file.
4. how to update the file for changes (file maintenance).
5. how to purge unwanted data from the file.
6. how to protect the file against loss.
7. how to relate this file to other files.

The first point must be positively resolved before doing anything about the others. There is no sense in figuring how to process, protect, etc., until you have made a realistic study as to what the file contents should be. No matter how efficiently you process something, you have not truly done an adequate job if you are processing something that shouldn't be processed, or if you are failing to provide something that should be provided.)

Should There Be a File?

Apparently the answer to this question is too often a quick and easy "yes." You probably know some homeowners who never throw anything away, and this same philosophy has been carried into many businesses.(If a business has grown up on the basis of just establishing more departments to perform specific steps rather than building systems around major applications, each department is prone to provide completely for itself.)With feeling the need to protect oneself, the desire to have information handy in one's own format (which may be slightly different from the wishes of those in the next department), and the feeling that most of the cost of a file is represented by the cost of the cabinet in which it is stored, (files may be too easily and loosely created.)

Of course, there may be legal requirements which necessitate maintaining certain records. This is especially true in an area such as payroll. But the typical business has so misinterpreted the requirements of the law that they are likely to go far beyond what is needed. For instance, a state law may require that payroll records be saved for three years. Some companies have understood this to mean that time clocks must be used to show time in and time out on a daily basis.(Actually, many businesses are now using an exception method of reporting time on the job; unless an exception form is filled out showing a person wasn't there or that he worked overtime, he is presumed to have been there and he gets paid for the normal work period.)This method of reporting time merely places a greater responsibility upon the supervisor of each area to maintain control over his employees and to make sure the necessary exception reports get prepared. If the rate at which exceptions occur is about 3 percent, try to visualize the reduction in processing steps and file requirements compared to preparing and keeping a positive record for everyone.

The only way to properly handle legal requirements is to get a meaningful interpretation of the state and federal laws involved and not to rely upon hearsay or what people may have assumed the law to be.

A large company formerly kept for many years all the correspondence relating to employee travel. Now they follow a procedure of reserving a hotel room in advance by letter or telegram with a request for confirmation by similar means. The employee about to make the trip holds the company's only copy of the request and the confirmation until he has made the trip; then he destroys both. The only record they keep regarding the transaction is the hotel receipt which shows everything that is needed. Any disagreement as to the rate charged should have been cleared up at the time of the visit.

Many companies have developed a way to reduce storage requirements by one-half and processing costs by considerably more on routine correspondence. Suppose an inquiry is received in regular letter form; the recipient merely writes the answer at the bottom of the letter in longhand, gets a picture copy of the letter, mails the copy to the requestor, and retains the original. (See Figure 5-1.) Some companies feel their "image" would be spoiled by following such a practice, and they stay with the method of typing an original reply. Actually, most people care more about a prompt reply than the form in which it is received. If you will accept as fact current studies which put the total costs of an average letter in the $1.50 to $2.50 range, you can see significant savings above and beyond those due just to filing procedures. Companies that have placed a high value on prestige factors often back off when they find that prestige may not be paying the bills or making a return for stockholders.

There was the case of a banker who saved all the envelopes in which his correspondence was received because twenty years ago he and the bank were instrumental in jailing a customer who had used the U.S. mails to defraud the bank. Apparently, he thought history might repeat itself. Can you imagine the file space taken up by the saving of twenty years' worth of envelopes?

Form in Which File Is Maintained

There are various forms in which records can be stored in a file. Examples would be: visual form on paper in filing cabinets, in the form of holes in punched cards, and electronic form in magnetic tape or disk. Paper tape, although used extensively for certain jobs in business, does not seem to be very attractive for most filing pur-

Figure 5-1.   Sample of a turnaround letter.

RICHARD W. LOTT
DATA PROCESSING AND FINANCIAL SERVICES
12 BEASOM STREET   NASHUA, N. H. 03060

(603) 882-3614

Mr. John Doe
1 Main Street
Hometown, U. S. A. 00000

Dear Mr. Doe:

    Please furnish me with a price list and normal delivery
schedule for a dozen of your brand X widgets.

    Thank you for your help.

                                    Sincerely,

                                    R W Lott

                                    R. W. Lott

Dear Sir:
Brand X widgets cost $45.00/
dozen. We can ship 3 days/
after we receive your order.

                                    John Doe

poses because of its relative bulkiness, slowness, susceptibility to breaking, difficulty to correct errors, and inability to be conveniently sorted.

In most cases, the method of storing will relate very closely to the basic processing methods of the company. If the company is small and has not done much in the way of office automation, chances are that storage will take the form of paper.

In the event the company uses punched cards extensively for processing, certain records will be maintained in punched card form in filing cabinets. In the inventory control application not only will cards be used to show the perpetual inventory for each item, detail cards that had been punched to process individual items will most likely be maintained to provide for possible future processing of those cards. The source documents (most likely in paper) on which the transactions originally were recorded would probably be saved because they represent a more formal record of the transaction in that they may contain authorization signatures and other units of data that never were key punched. Thus, this company probably has at least three separate files: the source documents, the detail cards, and the current balance cards.

If the company is using magnetic tape as its processing media, the detail records are likely to be kept in tape form in a vault. As in the punched card illustration above, the company probably has saved the source documents which were used to record the original transactions. Note that in an application such as inventory control where the volume may be extremely heavy, tape may not be the best storage media. This can be true in a case where on a given day the percentage of items in the file is relatively inactive, both from the standpoint of transactions that have occurred and need to be processed and from inquiries of the file that are to be answered. If only 5 percent of the items are active, the whole file must be run through the computer anyway in order to process the active items.

If the company uses the magnetic disk file approach, then certain records are going to be on small, removable disks in a vault or on a large disk permanently attached to the computer. However, most disk processing has been set up so that current balances and only major historical figures are maintained, and the detailed transactions are not on the disk. In the event that the detailed records of the past are required, the company may have them on magnetic tape, or if their computer does not have magnetic tape devices, the original input to the computer probably would have been in card form and the cards would be saved for a period of time. Again, the paper containing the original transactions would be saved.

In all cases, except with the disk, the files would have to be maintained sequentially or it must be possible to have them conveniently sequenced so that any processing significance could be established. In the case of the disk file, the items could either be stored in sequence or randomly if the programming had been set up to handle the items on a random basis. For reporting purposes, however, the file would have to be manipulated in such a way that certain sequential reports could be prepared.

The form of storage that is used will largely be determined by how fast processing is to occur and how much money the user wants to spend.

Method of Accessing

The problems involved in using the file must be considered at the same time you are considering the form in which the file must be kept. Before you start to consider the cost connected with any particular storage method, you should first be concerned with some general processing requirements. Two main considerations relate to the number of transactions and how quickly you need to process them. You may not have too much trouble estimating the number of transactions that are going to occur when your new system is operating, but do keep in mind some of the problems described on pages 218 and 219.

The first time you start asking operating people how fast they must get transactions processed you are likely to be told such things as: "Right away." "Within seconds." "As soon as possible." Another common thing you will hear is, "Maybe I can wait about a minute for processing, but I know John Doe in Department 4 needs it immediately." When you begin analyzing these claims and press on for more details, you often find that processing is not really required that quickly. There is plenty of evidence to show that costs rise rapidly as you try to speed things up, and these increases are often exponential instead of just straight-line progressions.

An example of the type of analysis you have to make is this: With what speed would a person have to be able to find out what an employee's year-to-date pay is? First, let's see some reasons why anyone would need this information.

1. At the time of preparing the regular payroll it is needed to determine whether certain taxes must still be applied to gross pay.
2. This information is needed at the time of printing W-2 tax return information at the end of the year.
3. When a person has been fired or is leaving without notice and is to be given a final pay check on the spot his year-to-date pay must be known.

In normal operations the first two procedures occur at a scheduled time and so do not require any quick action. Item 3 is the only one for which there may be a valid need for immediate access. But the value of providing such a service must be measured by the cost for a procedure which hopefully happens so seldom. The typical stumbling block to getting rid of such a procedure is that someone says, "It's our company policy to do so." Then it is your duty to point out to the proper people how costly this policy is in relation to its value. An alternative might be making out the check on the next regular processing day and mailing it or making an appointment for the person to pick it up. But if a strong labor union is involved, you may not be able to change the procedure.

In a situation such as the previous, you must try to find out why you need to have any procedure—that is, why so many people get fired or why so many people become dissatisfied and leave without giving notice.

How to Update for Changes

A file can properly serve its intended purpose only when it reflects current operating conditions. It is desirable to make changes in the file as those changes occur. For example, the pay rate or number of exemptions of an employee will change; perhaps a customer to whom your company has been selling has changed his address; or your management may have decided to change the name of one of your products. In order to make sure that future transactions are properly handled, the appropriate file records must be altered.

A business has to determine whether it needs to update the file as soon as a change occurs or whether it can afford to wait until a later time. A good reason for waiting is that you can economize somewhat by batching changes and making all of them for a recent period at one time. On the other hand, a file is always going to have the older data in it because of this waiting period. It can become irritating for a customer to mail a check to his vendor on October 18 and find that his account has not yet been credited by October 25 when the vendor makes up new statements. The important thing is to make changes prior to the time that regular transactions are next processed. If a person's rate of pay is increased and he is told of that fact on Monday, there does not seem to be a great rush to make the change in his file record so long as it is done by the time gross pay is next calculated.

The specific method by which changes are made will be determined mostly by the nature of the file. If the file is in paper form, quite often the change can be made easily by pulling out the sheet of paper, drawing a line through the old item, and merely writing in the change. Or it may be desirable in some cases to pull the sheet of paper from the file, write out or type a new one, and place the new updated sheet in its place.

In updating punched card files, you could manually pull out the card that requires the change, key punch a new one, insert it, and either save the old one or throw it away. If there are many changes to make in a card file, it might be most efficient to key-punch cards for all the changes, sort the cards to the order of the file, and use the collator to pull out the old cards and insert the new ones.

In magnetic tape you would need to keypunch a correction card or key the correction to tape, sort the changes to file sequence, and process them against the file that is already in sequence. This process would then write a new magnetic tape that would merely contain a copy of all the unchanged items and include the changed items in their sequential positions. On a disk file the process would most likely put the new item onto the same place on the disk where the old one existed.

In many cases it is desirable to later be able to see what the new file record is in relation to the old one. In the case of paper files, it is easy to see where the strike-out method was used. Otherwise it would be necessary to retain the sheets of paper pulled from the files. Punched cards can be saved to show how they appeared in that file before. In the case of magnetic tape, you would have the old tape that had been used to write the new tape. In the disk file, the only method of keeping track of the old transactions would be to periodically copy the disk to another and save one disk for later reference. In the case of cards, tape, and disk files, many people have found it desirable to print out the old record and also the new record at the time of updating and have somebody store those printed reports so that an audit could later be made of what had taken place.

How to Purge

A vital point in any data processing system relates to a well-planned and well-executed program of retaining data only for a stated period of time. Some requirements for retention are legal in nature, and others relate to the need to satisfy internal uses at a later time. The user must carefully select the form in which records will be kept, where they will be located, and how long they will be kept so that applicable costs are kept to a reasonable level. Obviously, much thought must be put into the specific functions that historical data must serve before making the arbitrary decision to keep all records for an indefinite period.

Many companies have allowed files to continue growing because they have not made a realistic effort to get rid of those items which no longer have any value. Another reason why a file can grow is because there may be transactions in it for items that in fact do not exist. For example, consider that a particular company does not have an item identified as part number 1743. But suppose the day comes when a sale is made that records a transaction for that part. If, as a result of that error the record is put into a file and if that entry is never reversed or pulled out, then that invalid item tags along with all the valid items. Furthermore, the item that should have a transaction shown for it does not.

The problems that inactive records create are basically three.

1. It takes longer to process active transactions in a file which contains items that are not active.
2. The cost of the file in terms of physically providing for it will be substantially higher if there are 20,000 items in it as opposed to having only 10,000 items in it.
3. With respect to files that are processed manually, more errors may be made because there are more items in the file. This increased error situation is not so likely to occur if the file is in one of the three automated versions.

Thus, it becomes a necessity for a business to maintain careful procedures for purging files at periodic times. For instance, they may find it desirable to clean out inactive items every six months or perhaps once a year. Surely there is someone within the organization who knows that a particular item is no longer likely to have any activity. Therefore, it can be pulled from the file.

A number of years ago one large company followed a practice of keeping about 70 percent of all its business records on a permanent basis. That company then put into effect a records management program so that now only 6 percent of its records are kept on a permanent basis. The company is following a program of taking a calculated risk that certain types of transactions will never need to be referred to, and many sales and purchase records of small amounts are being thrown away within one year after their preparation. The company is following the basic approach that storage can cost a great deal more than the value received.

They have set up a well-planned program to clearly indicate the life cycles of all types of records. One of the main points in their program is to make sure that every employee understands how the retention cycle is supposed to work. The company found one interesting thing that had been the cause of keeping many of the old records. If there were a law in existence that a certain form developed within an application had to be kept for ten years but some other form had to be kept only two years, there was a tendency to keep both forms for the full ten years.

It does become difficult to determine what records shall be thrown away. For example, a city in the suburbs of Boston has decided to use a computer to help figure out when a false fire alarm will next be turned in along with the alarm box on which it will occur. The Fire Department feeds false alarm statistics from the past several years into the computer and the program then predicts what the future will be like based upon the past. That operation has been very successful in catching offenders because such people seem to follow a certain pattern of turning in false alarms. What approach could that city possibly have used if they had thrown away all of those records shortly after the false alarm had been turned in?

It seems to be clear that management must help decide what type of retention program must be adopted. Thus, if records are kept, everyone concerned knows that it is going to cost money, and if it is decided to throw them away, no one person is going to be reprimanded because he made the decision himself.

How to Protect

If a file is worth having, then it is worth protecting against possible damage or complete loss. Such damage or destruction may come about from two main sources.

1. Employee errors of the programming type may occur whereby the computer might prematurely write on a tape or disk and destroy what is there. Or in operating a manual system a person may put the wrong throw-away date on a box of papers, and the box may be thrown away too soon.
2. Those types of disasters might occur which would physically destroy the file. Included are such things as fire, flood, theft, or intentional destruction caused by disgruntled employees or rioting outsiders.

The first type of risk can be best protected against by setting up good procedures and training people in their proper use. If procedures are simplified and if the right people can be selected for the various jobs, errors are not so likely to occur. Then if the proper self-checking and control techniques are used, many errors can be caught before extensive problems develop. Computer programmers must be instructed in the ways of designing certain steps into programs so that tapes can not be placed on the wrong tape drives, or that tapes are not used as output reels before they should be. This can be adequately handled by using proper header labels on tapes and following precise programming standards. Many of the software packages available today will provide for such steps without a great additional effort on the part of the company programmers. The measures described in the next four paragraphs will help a company to recover if an error is not caught soon enough to prevent a problem.

In the case of physical disasters which may occur, there are several precautions that may be taken to minimize their effect. In order to provide protection against a fire, most businesses will design very intricate vault systems so that magnetic tape and disk files are stored in the vault at all times other than when they are actually on their way to the computer room or being processed. Businesses will often install fire detectors in these facilities so that as soon as a slight trace of smoke is detected, the fire alarm would sound and alert everyone in the area. Of course, everyone must be trained as to what to do in such a case. It is interesting to note that magnetic tape does begin to decompose at approximately 180 degrees, so you can see that even the beginning of a minor fire will cause immediate damage to that storage media. An advertisement from one company shows that a fire in one office was hot enough to destroy many paper records in regular file cabinets but was not intense enough to burn a wooden chair.

2.   In the case of flooding, the best precaution to take is not to store any records in the basement or lower floors of a building within a reasonable area in which a flood may ever happen. One state had over one million punched cards completely covered by water in their basement facilities. When the water subsided only a fraction of the cards could be machine processed; the others had to be completely re-keypunched. Some companies will make sure they deactivate any automatic sprinkler systems in computer centers so that water does not damage computer equipment or computer files that may be in the area. But they must obviously have fire detection equipment and provide other means to put out a fire.

3.   In order to assure protection from sabotage or industrial spying, most businesses will try to provide adequate measures against unauthorized entry into the area where the files are stored. These latter measures should also apply to most employees of the business since they would have no need to enter the area. However, one risk that is very difficult to protect itself against is the disgruntled employee of the Computer Department who decides to leave the company and to damage files before doing so.

4.   Many businesses now have very precise procedures for periodically taking duplicate file records to remote locations. These duplicates are prepared by making copies of valuable papers, cards, tapes, and disks. Microfilm is also extensively used. The use of this procedure would tend to cut down the possibility that a major fire, flood, earthquake, nuclear explosion, etc., at the main building location would also occur at the remote site where the duplicate files were stored. Many businesses rent underground storage space for this purpose, and if something should happen to the main building location, the company would at least have its files as they existed at the time when copies were last taken to the remote spot. Some users do this daily and some do it perhaps once a week or once a month depending upon the nature of their operations and the volume of transactions and number of files.

In summary, file protection requires procedures to prevent loss and procedures to recover from loss should it occur.

File Relationships

Due to the fact that information for some applications may be needed by several departments, a system must be designed so that needed information is made available to everyone concerned without wasteful duplication.

This requires a careful study to determine if files should best be located in operating departments or in a central filing area.

Inventory Control

Several years ago the directors of a professional organization that had nationwide chapters established a committee to study what it would cost to set up a full-time executive director and related office staff. At that time the organization was being administered by volunteer help from the members with a slight expense for clerical help. After a lengthy study the committee submitted a detailed report to the board. About half way through the presentation, one person said, "Why do we need an executive director?" The whole room was strangely silent; in a few seconds everyone seemed to be asking the same questions, and not one person could give an adequate reason for having the position filled. Conceivably the organization may have hired a person for the job without ever having adequately determined a good need for him.

So before a file is designed, the need for it should be questioned. It is a good idea to consider the reasons for both a negative and a positive answer. Let's first consider the reasons why you would decide not to have any inventory files at all. This does not mean to imply you would not have any inventory. It means you might choose not to keep any detail records of that inventory. This approach is often used with success for certain types of small-value items or in those cases where the business may not be adversely affected by having less formal methods.

The common reason for keeping inventory control records is so that someone is constantly reviewing the current balance of an item, and when that balance reaches a predetermined point, a re-order is placed within your own Manufacturing Department or with an outside vendor to obtain more. In the case of an item such as pencils, paper, paper clips, etc., it is possible that a suitable vendor is close enough so that one-day or even hourly service might be obtained, and the person who took the last box might be the one to requisition more.

Some companies have been amazed to find that records in certain inventory files cost more to maintain than the value of the items they are controlling. One company found that for 25 percent of their items the cost of the related paperwork was greater than the value of the item. For one of their maintenance storerooms they performed calculations as follows:

$$\frac{\text{Cost of paperwork}}{\text{Number of requisitions}} = \frac{\$12,250}{20,000} = \begin{array}{l}\$0.6125\text{ processing}\\ \text{cost per}\\ \text{requisition}\end{array}$$

$$\frac{\text{Total value of material issued}}{\text{Number of requisitions}} = \frac{\$52,000}{20,000} = \begin{array}{l}\$2.60\text{ material value}\\ \text{per requisition}\end{array}$$

Since it took the same amount of effort to record a transaction for a bolt (value $0.25) as it did for a motor (value $52.00), the $0.6125 average was a realistic processing cost. Then they took a sample of the value of the material that was typically requisitioned; on one out of four requisitions the material value was less than $0.61.

They eventually eliminated all recordkeeping with respect to very low-value items and concentrated their filing system upon items which were critical to manufacturing and sales needs. They were also able to prevent the hiring of another person who would have been needed because of increased volume had they continued under the former method.

Some companies have used a very simple way to determine when it is time to reorder goods that are physically placed in the bin for sale or withdrawal for internal use. A particular mark, color, stripe, or similar indication is put on that box or boxes which indicates the minimum quantity or reorder point. When the person issuing the inventory notices he is down to that indicated item, then it is time to order more. Thus, any records of those transactions have been eliminated. Of course, it does require carefully selected people and close attention to operating details in order to survive with processing of this type.

Several factors supporting the maintenance of inventory control files are as follows.

— 1. By keeping detailed records you may be able to better control the theft or loss of items. By making people prepare paperwork and then properly controlling all the processing steps, you may be able to determine from a periodic physical count that the inventory on hand does not agree with the paperwork figure. This can be a tip-off that something is wrong, and by carefully reviewing the steps you may find the error. In the absence of any records, your basic assumption is that all the material has gone into regular manufacturing or sales channels.

— 2. A common reason for keeping a file regarding inventory control is to be able to conveniently prepare a financial figure for balance sheet purposes. If the file records contain both quantities on hand and unit costs, it is relatively easy to multiply and get a dollar value on hand for each item. The accumulated total can then represent the inventory figure on the balance sheet at statement time. However, if this is the only purpose for maintaining records on inventory, it represents a rather significant cost in relation to the value received. Consider that estimating techniques may provide reasonable figures for interim purposes. A physical count will be most likely used for the annual report, however.

3. Another reason for maintaining an inventory file is to be able to arrive at a cost-of-goods-sold figure for income statement purposes. If ten widgets with a value of $2 each have been sold, the Accounting Department can multiply quantity times cost and obtain a cost-of-goods-sold on this item of $20. This approach can be used to arrive at a total cost of all goods sold within any accounting period. However, if this is the only purpose for maintaining the files, then the business may be spending money for data which could be reasonably obtained by other means. One manufacturing company realized a considerable savings by using an average cost-of-goods approach. Their products fell into rather clearly defined product groups, and they determined that by merely multiplying the total sales dollar value for each group by its respective historical average cost relationship, they could in effect cost out what was sold. After such an approach had been used for one year, the company's public accounting firm made a substantial audit of two months' transactions. On a monthly cost-of-goods-sold base of $3 million they found a difference between the precise method and the average cost method of approximately $2000. The company knew that it would have cost much more than the $2000 per month to work out the cost on each individual sale, and the small percentage variation from actual was not significant enough to be of concern.

- 4. (Another very good reason for periodically looking at the inventory file from the paperwork standpoint is that greater control can be exercised over the people who have physical possession of the inventory) Operating people can become lax in their work over a period of time and perhaps continue to allow very slow-moving items to remain on hand long after it has become obvious that they are not worth keeping. So if management will see that systems are designed to make proper reviews of inventory matters, they may be able to uncover or prevent situations of this type.

Contents in the Files

The information contents of an inventory file can only be determined after you find out what it is that people want to keep track of. (Typical contents may be such things as follows: the identification code of the item, the name of the item, its unit cost, perhaps its normal selling price, number on hand, reorder point, reorder quantity, number sold in total last year, number sold in total two years ago; and how many have been sold from each sales office or warehouse.) Notice that the last few items do not directly relate to the inventory itself but rather to records regarding sales activity over the past. However, it does make a great deal of sense to have this information contained in this file and thus eliminate the need for having a duplicate file history at some other place. Figure 5-2 shows the format of the printout of one item from an inventory file at an industrial company. An inventory analyst at that com-

Figure 5-2. Sample inventory printout.

Jones Company   Inventory Report

Part Number	Description	Source		Lead Time, Weeks	Reorder Qty.	Reorder Point
		Bought	Mfg.			
12793	3" Widget	X		4	500	150

On Hand	On Order		Sales			
			This Month	Last Month	This Year	Last Year
132	500		512	436	3971	3684

Unit Cost	Date Introduced			Warehouse Location			Analyst	Today's Date		
				Bin	Row	Shelf				
1.74	5	18	64	12	A	5	Smith	10	27	2

pany can request a printout of any item he chooses. Depending on the form in which the file is maintained, an analyst may get the report in a few minutes or it may take several days.

Careful attention must be paid to each of the reasons there are for the existence of the file. Then if there must be a file, you need to make sure it contains the proper things so that all interests can be served.

Questions

1. Why do most systems require reference to files in order to process transactions?

2. Labor laws require that companies use time clocks to record payroll data. True or false?

3. List five different filing media. Which of the five is used the least? Why?

4. If a company uses a computer that has magnetic tape or disks, why will they still have a lot of records in paper form?

5. Give an example of a situation in which a random file would have to be printed out in sequence.

6. Suppose a company had a system that recorded sales transactions so quickly that complete details for a day were available by 8:30 the next morning. What could top management do with such information to better run the business if sales figures were ready that quickly?

7. What is file maintenance?

8. What can a company do to prevent the need to make daily changes to files containing labor rates?

9. For the following applications, indicate how often file maintenance should be performed:

   a. subscriber's address in a magazine company.

   b. deposits to and checks drawn on a checking account.

   c. number of dependents in a payroll operation.

   d. stockholder list in a corporation.

10. Suppose you were responsible for maintaining a file of the names and addresses of your vendors. What criteria would you use for deleting a vendor record from the file?

11. Make a list of the reasons why a company might keep records on inventory. Which reason is probably the most important?

12. In one year a company had $62,000 worth of material processed through its storeroom. It required 24,000 requisitions to handle that volume at a processing cost of $15,000. Suppose they changed their systems so that they could cut the number of requisitions in half and reduce the total processing cost by 30 percent. What would be the amount of the change in the processing cost per requisition? How would you justify the change in unit cost to management?

13. "Magnetic tape is better for payroll than it is for inventory control." Give a very sound reason why this might be true.

Other Readings

How Long Must a Taxpayer Keep His Records?, Massachusetts Society of CPA's, Inc., 1966. Furnishes a sound review of the problem of records destruction along with appropriate legal references.

"1969 Retention Timetable," *Modern Office Procedures*, April 1969, p. 24. This article gives a sound basis for retaining and disposing of most common business documents.

CHAPTER 3

CHAPTER 6

Studying the Present System

There is a continuing controversy regarding the topic of studying the present system. The basic questions relate to whether or not a study should be made, at what point in the whole system design cycle it should be performed if it is deemed necessary, and how much of a study should be made. There is no clear-cut answer as to which approach should be used in each specific case, but there are some guidelines which may be of general help. There are, no doubt, many analysts who have found a particular approach works best for them, and in some cases the systems supervisor or management will indicate the approach that the analyst is to take.

Why Study the Present System?

First, let's see why it might be desirable to make a study of the present system.

1. You are about to mechanize portions of the system. You will need to know what people do and how they do it so that you can program machines to assume those operations. If people are presently using a set of tables to figure state income tax, you must obtain those tables or appropriate formulaes to include in computer programs.

2. Perhaps a bottleneck has appeared in an operation. You must learn enough about the operation to be able to go in and eliminate the trouble. Perhaps reports are late, the error rate may be up, or efficiency may be falling off.

3. Because of a change in information requirements, perhaps some data must be prepared sooner or a new step or report must be added. You must know enough about the system to determine how and where to make the alterations. For instance, a state in which you do business may have just imposed a sales and use tax; you must institute means to determine those customers (and products) who are subject to the tax, and see that the proper amount gets invoiced, collected, and turned over to the appropriate government body.

4. Typically, the estimated cost of a proposed system will be compared to the actual cost of the present system. It is quite likely there is no one accounting figure available that indicates the cost of a present system, since most systems cross many departmental lines. Certainly the total costs of a "Payroll" Department don't include all the costs related to preparing payrolls. The Treasurer's

Office, Accounting Department, and the supervisors of every department all incur costs relative to payroll preparation. If the analyst takes the narrow-minded view and just looks at and works with a single department in his study, there will only be one area of costs included and he may have overlooked others that incur substantial costs.

5. A key employee in a department may have been in a disabling accident or no longer available because of retirement or resignation. Someone must come in and learn enough about the system so that it can be properly operated and supervised. Of course, it is desirable that a company always have adequate backup for situations like this, but it is not always available.

6. Any system that you help to design is going to be operated by people in user departments. You will have a much better chance of getting along with those people if you have worked closely with them. If you see what appears to be a good billing system in a magazine article and attempt to dump it upon the people in your Billing Department, you may have problems getting their cooperation. You must show that you have worked on their problems directly.

7. Those steps which will result in ineffective operations are not always obvious. You may be able to prevent the adoption of wrong steps by looking at present methods and then avoiding certain procedures in your new operations. One company completely relied upon sketchy information on customer purchase orders concerning the application of the state sales tax. Later the auditor for the state drew up a sample of sales made in his state and determined that the company was liable for an additional state sales tax of $10,000 for the previous year. The company was forced to pay the assessment; however, because of customer relations, management decided not to try to collect the applicable tax from their customers. Their new system included steps to make sure they collected sales tax according to the law, which meant they required a specific sales tax exemption form to be received with the customer's order when no tax was to apply.

8. A study of a system will also help to determine if a supervisor really has any justification for adding people to his staff or to perhaps determine if a budget request is reasonable. One systems analyst made a brief review of a department when it placed a requisition for two more people. The analyst saw a place where there was great duplication of effort and recommended that two of the existing people be declared excess.

Conversely, there may be some reasons for not studying the present system.

1. If management and operating people are not satisfied with the present system, it may be better if you don't question it too closely. Thus, if you don't study it, you will gain no biases. When the Budget Rent-A-Car division of Transamerica Corporation began franchising rental agencies, it did so with the stipulation that no franchisee have previous car rental experience. They did this so they would be sure to get people who "didn't know it couldn't be done." In the few instances where the company made exceptions to the rule, the results were always disappointing. If you know nothing about the details of a system, then you won't be automatically copying any of its bad aspects. At least if the wrong things creep in, you will have arrived at them independently.

2. An employer may be willing to spend for research and development purposes. One large company with sixteen similar manufacturing plants installed random access computers in fifteen plants and a magnetic tape computer in the other one. They did so in order to compare random access to sequential processing after a year of operation. For this reason the Systems Department at the sixteenth plant was not given any opportunity to find out what the other plants were doing.

So you can see there are valid reasons for taking the approach of whether or not to study. The entire systems group should carefully determine whether it should go ahead with a study or not.

Occasionally an organization will enter into a completely new activity as far as it is concerned, and there will be no present system in operation to study. The J.C. Penney Company department store chain had this situation in the early 1960's when they decided to switch from a strictly cash basis to one of providing charge accounts. They weren't going to just adopt a new way of doing something—they were about to do something they had not done before.

In a situation like this it is usually possible to learn extensively from your competition. At least it is better to try to obtain help from them and get turned down rather than blindly assume they wouldn't help at all. Organizations are typically proud of some of the things they have done, and many will provide information through meetings and publications of groups like Administrative Management Society, Data Processing Management Association, and Association for Systems Management.

A friend of mine was once part of a management group of a

steel company who spent six months with a foreign government to help them set up a steel plant. Six months after the plant was in operation, it was selling steel in the United States and competing with the company who had helped to build it. The domestic company did not regret what it had done because if it hadn't, some other steel company would have, and the earnings from the consulting work had made a significant contribution to profits.

What Data to Study

(Once the decision has been made to study the present system, the analyst must determine what it is he is looking for or specifically wants to get out of the study. In fact, this should be the major consideration in determining whether to make a study. If he isn't looking for something specific, then maybe he doesn't have a sound enough reason to go ahead.)

During his first day on a particular systems assignment, an analyst was told to go to the supervisor of the Accounts Receivable Department and find out how his system worked. When he asked what he was to get, the boss said "everything." Later on he found out why this was the case. The whole study was being made that way because it was not organized at all and everyone was actually groping around in the hope that some major idea would eventually hit all of the analysts.

When the analyst arranged to talk with the Accounts Receivable supervisor, he was told he was now the *fourth* person in two months who came looking for "everything." Even the systems analysis project leaders weren't sure about what the other was doing, and they had made no attempt to coordinate their efforts and/or share the information the various groups uncovered. Not only was the study of the present system too expensive because of duplication, but they also spent time uncovering information they did not need to know. Also, the people in the department under study wondered how effective the Systems Department could be if they conducted themselves in such a disorganized manner all the time.

(The following is a list of the types of things you might want to obtain and some possible reasons for collecting them.)

1. Samples of all input forms used must be looked at. This will help you to <u>find out what the raw data is</u>. By getting copies

of real input forms that have already been filled out you can determine the quality of transaction recording.

2. What is the flow of forms through the various departments; find out where things come from, where they go, and what is done to them.

3. See samples of all reports prepared, a way of finding out what is done with the raw data and how it is manipulated.

4. Samples of all reports received are helpful as a double-check on the reports prepared in 3 above. Is something else happening that you haven't learned about? It is if a department receives a report that you did not know was being prepared.

5. Volume of transactions processed and their likely growth rate must be known in order to plan for equipment capacity and number of people needed. It is necessary to gather accurate statistics rather than rely upon guess work or intuition. In one company many people in a meeting were trying to figure out the percentage of customer invoices which contained more than one item. Estimates ranged as high as 50 percent, with no estimates less than 10 percent. Finally, two volunteers took the time to pull 2000 invoices at random from the files. Everyone was shocked when they found that only five of the invoices (5 invoices/2000 invoices = 0.25%) contained more than one item.

6. The number of people per section, number of file cabinets, and number and types of mechanical devices can be used to help arrive at present system costs. It may also have possible use in determining the output you can expect from the personnel in each area.

7. You may be in a position to determine who the best people are for certain jobs. Or you may be asked to comment upon those who aren't qualified to continue in the new operation.

8. What information and qualifications must people have to make certain decisions? Can the rules used to make decisions be automated?

9. Determine a rough breakdown of people according to general duties. You may find that ten different people all do some filing but in total it does not amount to eight hours of work per day. If filing could be automated, you would not be able to eliminate one person because of that.

10. Conduct personal interviews with workers to find out if they know their true functions and where they fit in the overall plan. Perhaps a proper training program will help improve operations and instill cooperation among departments.

11. During personal interviews with workers find out if they have suggestions for improvements. The people working in the department know more about it than anyone else and they should have the opportunity to be heard.

12. Find out the odd things that have happened so that the new design is as flexible as possible to anticipate them and provide reasonable steps to handle them.

13. Obtain a general idea as to how satisfied people are. If people are satisfied you can often conclude you have good systems. At least this is a better sign than when complaints and high turnover prevail.

How to Make the Study

Preparing to study the present system gets down to the nitty-gritty details of digging out facts. Certainly the first thing to do is to tell all operating people exactly why the study is being made. This disclosure will eliminate all the undesirable rumors that otherwise come up, and you are likely to get better cooperation from those whose help you are soliciting.

In most cases, you will uncover the information you seek as a result of personal interviews with the people who are doing the work. You certainly should not rely upon getting all the details from the supervisor because he may not be that familiar with details and he may possibly have reasons for telling you only what he wants you to hear.

Another very useful means of obtaining the facts is by actually going to the department and observing actions. Some analysts have actually taken up residence at a desk in the area involved. Another possible way is to study available flow charts and operating manuals on how the system is operated. But the problem with this method is that so often the system is not operating the way the manual shows due to changes which have never been updated in the manual.

In addition to gathering the facts, it will be useful to find out why certain things are done the way they are. This will merely help you to understand the overall system better, and it may keep you from making mistakes that you would otherwise make if you weren't familiar with the operation.

The analyst obviously should not rely upon his memory to retain all the information he has gathered. Most people will find it necessary to write down what they have learned as they get it. This can usually be done in the person's own "shorthand" at the time of first learning it, but it should be transcribed into some permanent, standard form within a few hours. A permanent record in narrative form might work but such an approach takes a lot of space and is relatively more difficult to read and digest as opposed to some version of flow chart. Since details gathered from the study of a system will normally be kept for some period of time and likely be reviewed by many other people, it is desirable that an organization adopt a standardized flow charting scheme.

For the purpose of this book the following general flow charting symbols will be used.

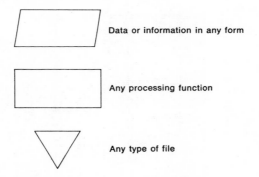

Data or information in any form

Any processing function

Any type of file

An example which shows how those symbols can be put together to show the flow of data in a system is shown later in the chapter. While flow charts may be ideal means to illustrate certain procedures, they become unwieldy in preparation and comprehension in cases where a lot of things are done to many different types of data. Not only should the recording method be easy for the analyst to use, it is also desirable to use a form that is understandable to operating personnel. Once the flow of data as you understand it has been prepared, you can show it to them. They may be able to suggest changes which would make it more complete because of their intimate knowledge of their system. If the recording method is simple enough, you might even get operating personnel to prepare it.

A recently developed tool that can be used to great advantage is the decision table. Although the decision table was originally designed to help prepare the internal logic of computer programs, it can serve equally well as a form of systems flow charting. You can easily construct your own decision table form by taking a blank sheet of paper and drawing lines for columns and rows similar to that in Figure 6-1.

Figure 6-1.   Form for a decision table.

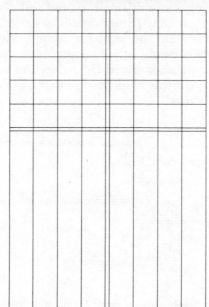

Suppose the system you were studying was related to the Purchasing Department. After clerical people type purchase orders (contracts that will be mailed to vendors to order products and services), the orders must be routed to specific personnel for their signatures as authorizations to buy. In your interviews with employees in the Purchasing Department, you find the following rules in effect.

1. Every buyer must sign all purchase orders he initiates. Each order has a code that identifies the buyer. He is solely responsible up to $500.
2. The manager of the Purchasing Department must countersign any orders up to $20,000, unless the order is made to a subsidiary. An exception to this rule is when a $501–$20,000 order is for capital equipment, in which case the General Manager must countersign.

3. All orders over $50,000 must be signed by the president, while those from $20,000 to $50,000 must go to the vice-president of Operations.

Figure 6-2. Decision table showing purchasing rules.

	1	2	3	4	5	6	All Other
Is order up to $500?	Y	N	N	N	N	N	
Is order $501–$20,000?	—	Y	Y	Y	N	N	
Is order to a subsidiary?	—	Y	Y	N	—	—	
Is order for capital equipment?	—	Y	N	N	—	—	
Is order $20,001–$50,000?	—	—	—	—	Y	N	
Is order over $50,000?	—	—	—	—	—	Y	
Buyer	X	X	X	X	X	X	
Manager, Purchasing Dept.				X			
General Manager		X					
V.P., Operations					X		
President						X	
See supervisor							X

A decision table which shows those present rules is shown in Figure 6-2. Notice that the table provides generally for the following.

1. Questions to be asked are in the upper left section. You may form the questions in such a way that answers will be yes or no.
2. Those questions are answered yes, no, or not applicable (—) in the upper right.
3. The various possible actions are written in the lower left section, one action per line.
4. X's are used in the lower right to match up actions with the criteria contained in the upper half of the table.

There are few helpful hints which will make preparing decision tables easy because they are so simple to begin with. One thing you would eventually develop is a technique to keep the table free of excess notations. For instance, Figure 6-3 tries to do exactly the same job as Figure 6-2 but shows a lot more activity only because of the order in which the questions are asked. Another feature which is desirable to show is what to do with "all other" situations; notice how the columns have been set up so that the last one provides for all other situations.

After you had prepared the table, it would be an easy matter to review it with purchasing personnel. This could be used as a means to get them to agree that it is accurate and complete. Or it may induce them to think of corrections that should be made.

Figure 6-3. Decision table showing purchasing rules (incomplete).

	1	2	3	4	5	6	7
Is order for capital equipment?	Y	Y	Y	Y	Y	Y	
Is order to a subsidiary?	Y	Y	N	Y	Y	N	
Is order up to $500?	Y	N	Y	N	N	N	
Is order $501–$20,000?	—	Y	—	N	N	Y	
Is order $20,001–$50,000?	—	—	—	Y	N	—	
Is order over $50,000?	—	—	—	—	Y	—	
Buyer	X	X	X	X	X	X	
Manager, Purchasing Dept.							
General Manager		X				X	
V.P., Operations				X			
President					X		
See supervisor							

Inventory Control

In beginning to study the present inventory control system you must not spend too much time studying details which are not going to be of value to you. As previously stated, it is desirable to study the present system to find out only those things which will be needed for later steps. Spending the minimum amount of time possible on this effort will give you more time to spend on analyzing what you have obtained and designing the necessary changes.

Certainly one of the major things you will want to study is just how inventory is physically handled. In this connection it would be desirable not to spend all your time at the desk in your office but to go out to the places where inventory is stored and moved about. This would probably involve a visit to the Receiving Department where goods are physically received from transportation companies or from your own manufacturing departments. From there you might want to trace its physical flow into bin storage or warehouse locations and at the same time observe as much as you can the paperwork procedures that are performed. At the same time you would want to check to see how the goods are physically pulled from those bins or warehouse locations and roughly what happens in order to get them ready for shipment or transferral to the next operation. In this connection, it might be desirable to spend several days in the area where inventory is physically handled because only after you see what physically happens to the product can you best determine the paperwork procedures that are necessary to control it.

During this process of observing the actions upon the physical inventory you will want to be very observant regarding the roles of the various people, both employees and people from outside concerns, to see what they are doing, how long it is taking them to do it, and so on. Any inventory control method you set up must be properly geared to handle these transactions within a reasonable amount of time so you don't have a lot of trucks backed up at your dock waiting to be loaded. At this time, you will get a rough idea of the general working conditions and the type of people who work there. You must be careful to design systems which do not overextend what you have to work with.

Regardless of all the other things that people may do in inventory control, one item of information that will be indispensible to you will be the number of transactions. Although your new system may do something which would alter that volume, certainly management has in mind that you will devise a system to take care of the present volume and the growth and normal fluctuations that are going to take place in it in the reasonably near future. You must be careful that you do not rely upon somebody's opinion regarding the number of transactions. You can best determine this by looking through prior records and finding out the volume in this manner.

You must not be mislead into counting or obtaining only an average volume because an average tends to overlook the very slack periods and also the peak periods. Obviously you must be able to provide for the processing of transactions at the peak periods because inventory control is not something you can allow to line up for several weeks before getting the transactions processed.

In order to keep the book to a reasonable length, we can only afford to study a few aspects of inventory control. For that reason, discussion will include only the following:

1. general flow of data involved in filling a customer order.
2. general flow of data involved in replenishing stock.
3. review of the individual item inventory control record.
4. summary of some of the month-end reports that are prepared.

Figure 6-4 shows the general flow of data in the processing of a customer order; the bottom section shows what the Inventory Control Department does to replenish stock. Figure 6-5 shows the general format of an individual item inventory control record along with transactions on that item for two months. The comments at the right side appear only so you can understand how the posting occurred.

Details of the inventory reports prepared at month-end are covered in Chapter 7 at the point where they are analyzed.

Questions

1. In which situation is there a greater need for study of the present system—when you are making systems changes or when you are basically mechanizing the system the way it exists?

2. Why is it difficult to determine the actual operating cost of a system?

3. Suppose you were designing a system to handle charge accounts for a department store. When you are at the point of estimating personnel and equipment requirements, what would be wrong with obtaining and using the number of transactions in an average month?

4. Since the introduction of computers, a lot of systems effort has been spent on studying the present system in detail. Why is this? Is this situation likely to change?

Figure 6-4. Flow chart of portion of inventory control system.

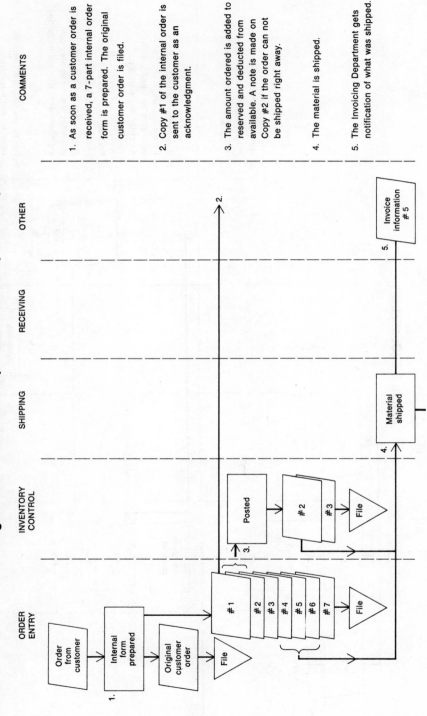

COMMENTS

1. As soon as a customer order is received, a 7-part internal order form is prepared. The original customer order is filed.

2. Copy #1 of the internal order is sent to the customer as an acknowledgment.

3. The amount ordered is added to reserved and deducted from available. A note is made on Copy #2 if the order can not be shipped right away.

4. The material is shipped.

5. The Invoicing Department gets notification of what was shipped.

Figure 6-4 (Continued)

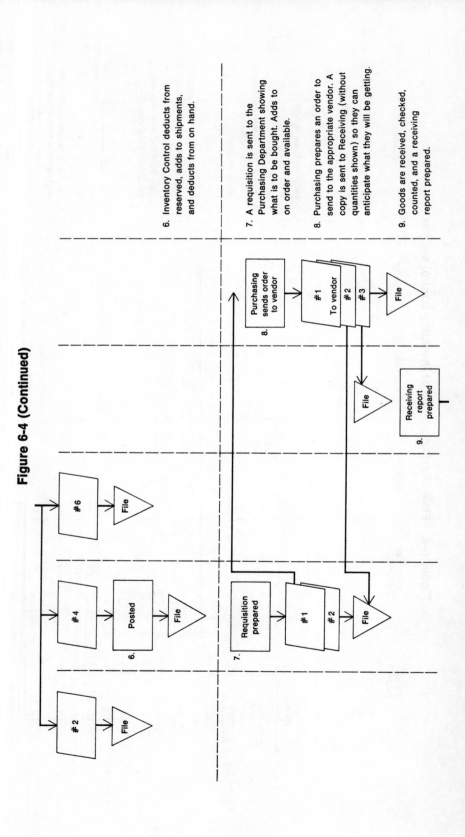

6. Inventory Control deducts from reserved, adds to shipments, and deducts from on hand.

7. A requisition is sent to the Purchasing Department showing what is to be bought. Adds to on order and available.

8. Purchasing prepares an order to send to the appropriate vendor. A copy is sent to Receiving (without quantities shown) so they can anticipate what they will be getting.

9. Goods are received, checked, counted, and a receiving report prepared.

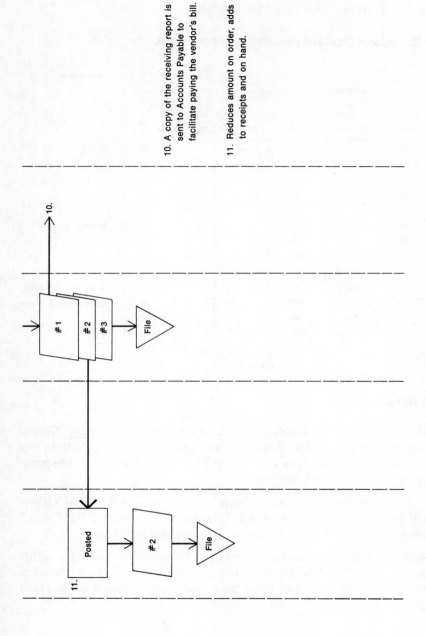

10. A copy of the receiving report is sent to Accounts Payable to facilitate paying the vendor's bill.

11. Reduces amount on order, adds to receipts and on hand.

Figure 6-5.   Jones Company inventory ledger card.

Comments

Part # 12794	Description Wankels		Reorder Pt. 25		Reorder Qty. 100	
Date	Reserved	On Order	Available	Receipts	Shipments	On Hand
1/1	50	100	75	—	—	25
1/2	50		75	100		125
1/6			75		50	75
1/7	44		31			75
1/12			31		44	31
1/17	23		8			31
1/18	23	100	108			31
1/21		100	108		23	8
1/24	42	100	66			8
1/29		100	108			8
1/31	300	100	(192)			8
2/4	300	400	108			8
2/16	300		108	400		408
2/19			108		300	108

Comments column:
- 1/17: Reorder required
- 1/29: 1/24 order cancelled
- 1/31: Extra required reorder

Other Readings

Nadler, Gerald, "Is Analysis of the Present System Really Necessary?", *Ideas for Management,* 1966, pp. 203–209. This article is a very logical approach to determining exactly what it is in the present system you need to study.

Neuschel, Richard F., *Management by System.* New York: McGraw-Hill Book Company, Inc., 1960. Chapter 9, "Gathering the Facts," gives dozens of ideas on studying the present system.

Sullivan, John W., "Requirements for Long-Range Systems Information Planning," *Ideas for Management,* 1966. Page 7 contains statements suggesting at what point you might ignore the present system.

CHAPTER 7

CHAPTER 7

General System Design

During the process of studying the present system, you will surely be getting design ideas for a new system in those places where changes are clearly required. But the design of a system so intricately involves many things you should not allow yourself to get preconceived ideas that you wouldn't be willing to change at a later time. It is necessary to put everything into its proper perspective through an analysis of what has been found.

Analysis of Findings

Once the present system has been studied in the way generally described in Chapter 6, you must take the time to make a careful analysis of what has been learned. You will probably find that your concentration at first should be related to what the general outputs of the system are; you should try to determine how adequately those outputs are serving the needs of the company. It is still much too early to think about how much the system is costing you or to be concerned with the details of performing any particular step. (It is not of much value to calculate the theoretical gasoline consumption of a car until such time as you know that it will run.)

Much of the analysis is going to directly involve the various people and departments in the company who make use of the outputs. Most likely you are not properly qualified to determine whether every segment of the operation is doing its job adequately, but there may be situations obvious to you where something is being done incorrectly, the output is not accomplished on time, or the system is just not serving its basic purpose. While you were studying the present system and gathering the pertinent facts, you surely uncovered material or personal comments that indicate inadequacies in the operation. This information can be a foundation on which to continue your analysis.

At this point it would be good to deal personally and directly with the various people and departments that would seem to have a "need to know" regarding the application you are studying. You must constantly keep referring to the objectives of the system as pointed out in Chapter 2 to make sure that you uncover all the places where this information is to be used.

It is also a good time to begin thinking of any possible advantages a meeting might have in which you would pull together various managers of the different departments involved. The purpose of such a meeting is generally reporting what has been found, restating company and department objectives, and getting general suggestions for improvement. It is also possible to learn of operating problems you or others may not be aware of. Hopefully everyone present will have a better understanding of the overall picture, and cooperation is more likely to occur as a result of the time taken.

Of course, if such a meeting should take place, you must do whatever is necessary to prevent it from breaking down into something negative if any problems are encountered. This gives further strength to the need for management to be informed and to make sure they give whatever support is necessary. For this reason you may want to insist that some top management be present at any meetings to help make sure that all your efforts are not in vain; people are more likely to cooperate and get something accomplished when management is present. A very important technique to follow at a meeting of this type is to record all comments whether they are of a constructive or destructive nature; but in most cases it is wise not to record the name of the person who makes the comments. Over a period of time, participants at such meetings will see they can speak their minds without later being embarrassed by published memos pointing them out. Since you as the coordinator of the meeting will be very busy asking questions, you won't have time to take notes and should probably have someone do it for you.

After the meeting or whatever other means has been used to gather all the necessary facts, you are ready to sit down again by yourself and try to pull all the pieces together. The Inventory Control section in this chapter illustrates how this might be done.

General System Design

Only after the basic nature of output reports and file contents have been established and the present system has been analyzed is it appropriate to begin the design of new processing steps and data flow. These prior steps are necessary to establish a solid framework. Any other approach is likely to result in too many compromises in what people will get since it becomes very easy to water down output by cutting corners here and there on processing. Or there may be a tendency to accept many of the inadequate things that are being done now. Once you get into the design of details and see how difficult or costly something is going to be is the time to consider backing off, not when you are in the broad design stage.

In so many cases the officials of a company will have obtained a computer or other specific piece of equipment, and the Data Processing Department will be required to begin using that equipment as soon as possible in operations. In such cases, the systems analyst may never have had a great deal of choice regarding the selection of the proper equipment but he may have had to alter the design of his system to fit the equipment. But the following presentation will describe what would be the ultimate situation; that is, the analyst has at his disposal whatever devices he needs to best operate each system he designs. Even for the typical analyst who obviously does not completely enjoy that situation, it would be worthwhile to become familiar with the approach presented in the next few pages.

You must not yet begin to think of any pieces of equipment or special gear that you would use in operating the system. You must concentrate only on the things that are needed in the way of output information, the necessary raw input, and the very general processing steps that will be applied to make the system operate. You do not need to provide for the last detail in planning for that processing.

At this point you are not deeply concerned with all of the exceptions that can occur to the normal stream of items, but you certainly would not avoid recording or thinking of any exceptions that happened to be brought to your attention or that you may be able to figure out on your own. As information regarding these exceptional items comes to your attention, you may begin to plan generally on having more than one main stream in your general system flow. During April and May 1968, operating employees of a large telephone company were on strike. Supervisory people were handling the regular duties of telephone operators, and they had so many problems just placing calls they weren't able to do the things necessary to charge overtime on long distance toll booth calls. In any future planning, the company would have to decide whether they should design the procedures that would assure the collection of such revenue.

At the same time as you are working on this general approach, you must be very aware of the types of controls that are going to be necessary in the system. You can probably get some ideas based upon controls which might be used or are lacking now. As soon as your work progresses to the point where a general system is taking shape, it will be necessary or at least highly desirable to confer with company and internal auditors to begin planning the controls that are going to be used. You must recognize that the system and its controls are not going to be designed for the auditors but for the users of the information that comes from it. But it is necessary to realize that any system is subject to later scrutiny by various auditing groups and management, and it is much better to plan for reasonable control ahead of time rather than too late. This can be done by enlisting their help early rather than to install something and then find that you are severely criticized or that the system breaks down because of the lack of control.

At this point you will also begin to think of the types of major things that can go wrong with the operation and to try to plan on the types of backup files and plan for recovery procedures that might be put into effect in the event that some major disaster should occur. You should not be concerned about all the various minor things that could go wrong with the operation of the system. This is only because you don't want to spend the time on those details which may not ever come about if your forthcoming recommendation is not accepted. You might generally approach this problem by answering the question, "What would happen if?" The broad concept you are concerned with is illustrated as follows.

1. A number of companies faced considerable problems when the Suez Canal was closed in 1967. They had to quickly find an alternative route for their ships.
2. Businesses who had closely estimated what profits would be and then committed themselves based upon those estimates may have run into considerable difficulty when the 10 percent income tax surcharge was put into effect in 1967.
3. Companies who have not been constantly on the lookout for alternative business have been hurt because of large contracts being cancelled or not renewed.
4. A major power failure in 1965 caused many companies to install emergency generating equipment.

Role of Transactions

Most data processing activities relate to doing something with transactions that have occurred, a transaction being one of the following.

1. A physical occurrence such as receiving goods from a vendor's truck into your warehouse or withdrawing material

from your storeroom to be placed into a production stream is a transaction.

2. A transaction can also be the receiving of information from an event such as opening the mail and finding a customer order. Or the situation where a worker has just completed an eight-hour work day and is expected to record (report) that fact so he can eventually be paid and the company can also charge that expense to the cost of doing business is also a transaction.

3. A transaction may take the form of a request for information. This could be an inquiry about a transaction that has already occurred or a request that will create a future transaction. The question could be, "Do we have enough stock to fill an order for part number 14835 now?"

From the standpoint of designing the ways that a system shall be operated once the outputs and inputs have been specified, it is necessary to review the broad types of systems. Whichever is chosen is largely a matter of how fast transactions must be processed and how much the user is willing to spend to accomplish it. These are the broad types of systems.

1. Batch, off line
   (a) *Collecting data*—Transactions are recorded independent of any computer control ("off line" means not attached to nor under the control of). They are "saved up" for a period of time and then processed later as a group. An example of this is a business might save all checks received from customers until 11:00 A.M. each day and then process them as a group to relieve accounts receivable of those customers who are paying their bills and to make one bank deposit for the day.
   (b) *Reporting results*—After a certain time period's worth of transactions have been collected, they are processed and reports are printed and distributed. The reports can be and often are voluminous. A common example is the printing of pay checks for all employees at one time in a continuous run.
2. Batch, on line
   (a) *Collecting data*—Transactions are recorded into devices in such a way that they are immediately under computer program control, and the data is usually

stored in magnetic tape or disk form. But processing is done later on a batch basis. This was merely a convenient way to get transactions into computer form without the common process of recording first on paper and then on keypunching cards. It can also be very effective to have the computer do some editing at recording time to detect and report obvious errors. Such a process requires on line files, tape or disk, to save the transactions since core storage is too expensive to use that way.

(b) *Reporting results*—People with the need to know have voice answer-back telephones (audio response), CRT screens, or printing devices to get an immediate answer to a request. However, the answers available are only as current as the processing of the last batch, which may have been last night, a week ago, or a month ago. In this style the requester can keep expecting to get the same answer on the same item all day long unless processing occurs more than once a day.

3. On line, real time

Collecting data and reporting results—Transactions are entered and processed as they occur, and results are immediately available to anyone who has been provided the means to obtain such output. This could be the one who furnished the transaction or it might be a person who merely uses the data supplied by someone else.

Based upon all the work you must have done to get this far in systems design, you should generally know whether a real time or a batch system is needed. From a historical point those businesses which have installed real time processing are typified by the following.

1. Airlines must know constant status of all flight seat availability in order to minimize the number of empty seats at take-off time.

2. Banks with many savings accounts may have a depositor come in at any time to make a withdrawal. The bank wants to service the account quickly and not cause an undue wait by the customer.

3. Telephone companies want to be able to answer your questions about your account immediately. Considering the

business they are in, a letter reply and its associated delay would hardly be appropriate.

Even though there are many companies who have individuals that believe real time processing is a necessity, only a very small percentage of businesses has gone that far. When a company reviews the tremendous resources needed to operate that way, they quite often realize they must be satisfied with something less costly and less timely.

Perhaps somewhere along the line, management has indicated a rather general "ball park" figure they are willing to spend on a system. Most every company will probably have some value in mind beyond which they will not go even if increased expenditure would provide something better. Management will recognize that you may tend to increase the value of any output by spending more on the system but for reasons relating to the law of diminishing return and the maximum amount of money available for a system, there will be some point which they will not exceed.

Since reports that are rendered and files that are maintained are the results of transactions that have occurred, a great deal of time must be spent to make sure that transactions get recorded properly.

There are three basic criteria to be used in determining the best method for recording.

1. The physical means which are then going to be used to process those transactions to the desired output must be considered. Is a computer with magnetic tape going to be the processing device?
2. The physical environment surrounding the place where the transactions occur is important. Is the recording location outside where it may become cold and wet or inside in a well-lighted office?
3. The ability of the typical person who is going to be performing the recording function is also important. Is the employee likely to be a well-trained college graduate or someone who probably hasn't completed high school?

In a general sense it can be stated that you must strive to get transactions into the form in which they will eventually be processed as soon as you possibly can. If the system is completely manual, there are obviously no alternatives except to build it around "paper." But if machines will be available, then you must try to cut as many of the manual steps as possible. This will obviously cut the number of places involving handling so there are fewer places where costs can be generated, fewer places where errors can occur, and the final processing can begin that much sooner.

In most cases there is more time spent recording transactions than upon any other phase of operating a system, and you must be very careful about the means chosen to do this. It is not sufficient to merely look at catalogs or call in vendors to see what recording devices are available, since by that method you would only see what certain vendors have. The better way might be to devise a new method of recording without being shackled by what you already know or by any other biases. You should also observe the area where the transactions are going to occur and also get ideas from the operating people since they are the ones who are going to be doing the work. With their expertise regarding the nature of the data and yours on data collection methods and devices, the work should get completed in the best way possible.

In most cases a transaction represents a two-sided situation, and thus two purposes are being served by the recording. The very nature of double-entry bookkeeping is that a transaction affects at least two accounts. Thus, everything must be carefully planned and executed so that all parties to a transaction are adequately covered. Here are some examples of this two-sided effect.

1. A sale in a department store involves information relating to the nature of the item sold for inventory control purposes and also involves the process of collecting from the customer (either in cash or obtaining his promise to pay represented by his credit approval).
2. An employee reporting time on a job represents a labor cost to a particular budget area and also the amount of money that must be computed and paid to him.
3. Collection of a bill by a company adds so much to a cash account and reduces an account receivable by that same amount.

In planning for the most effective way of recording transactions, it is necessary to consider circumstances such as the following.

1. Is your first knowledge of the transaction the result of a phone call, a letter, or a face-to-face meeting with an outsider?

2. Is it a one-way transaction? That is, does information go only toward the system such as it might where workers on the job send raw data to the Payroll Department without anything necessarily going back to that area? Or must the system provide a physical object such as material or data such as a receipt to the person sending the original data?

3. What is a reasonable amount of time in which to complete the transaction? A person standing in line expects an answer sooner than one who has written a letter.

4. How much of a loss are you subjected to if an error has been made on your side of the transaction or is there a reasonable means to catch errors before they create losses? What kinds of errors can be made? If an airline makes an error on seat count and the plane leaves with space available, when there was someone who wanted that flight, they may never be able to recover that lost sale.

5. Is it likely that the system will ever completely break down? A business can never know when a disaster will strike and thus completely disrupt their plans. In January 1967, McCormick Place in Chicago, at that time the nation's largest convention center, burned to the ground. The building had been known to be "fireproof." But despite that, the whole structure was wiped out along with about $100 million worth of equipment on exhibit at the time. Many companies had taken their only prototype of a new product and had it on display. The insurance premium on the building was only $20,750 a year because of the reputed nature of its construction. One insurance company estimated they would have to insure a place of that type for 1500 years without any claims in order to make up for the loss that was suffered on this single event. This illustration just points out how carefully you must provide for everything because something not under your control could completely wipe you out.

6. Will the volume at times get so great as to completely overwhelm you as has happened in stock brokerage and utility firms? (Notice the special pricing deals that airlines and telephone companies have developed to spread volume out over a wider period and thus eliminate peaks and valleys of activity.) Or will volume possibly drop off to the point where fixed costs of operations will be spread over too thin a base? The high fixed cost of operation has made it most difficult for a low-volume automobile manufacturer to stay in business.

7. Is there a better basic way for the transaction to take place? (For instance, can you induce people to order by mail rather than in person?) You want to do things well, but there is no point in doing something well that which should be done in a completely different way.

After you have become completely satisfied that your general system design provides for the information which has been specified and after you determine the input which is necessary to provide for that basic information, you are ready to begin a preliminary study of whatever types of equipment might be needed to marry the output to the input. For this purpose a knowledge of available equipment will be very helpful. But it will also be most helpful if you will prescribe the specifications of your requirements when you find that a particular machine is needed but is not available. These specifications can be circulated to various vendors to see if such a device can be designed and roughly what the cost would be. Many of the pieces of equipment on the market today came about because of that very approach by far-sighted managers who were not willing to be content with less than ideal equipment that was already on the market. When American Airlines was designing their on line, real time system for their airplane reservations in the mid-1950's, there were various devices they realized would be needed to operate the system so they went out to various vendors and eventually had the devices built to their specifications. Of course, a small company can only go so far to develop equipment on this basis, and they will usually have to be satisfied with some of the equipment available. However, they must recognize that by spending a certain amount of money on research and travel, they may be able to find equipment suited to their needs as opposed perhaps to only casually looking through literature or calling in one or two vendors.

Fresh Approaches to Systems Design

Many of the systems in operation today do not resemble their counterparts of the past. As a result of the need to develop ways of generating more profits and often because of the influx of new, young people, significant changes have been made. The next few pages illustrate new ways of meeting certain objectives that have been a real break from tradition.

Many years ago Eastern Airlines began operating a "shuttle" airline service between Boston and New York and between New York and Washington, D.C. The system operates substantially in this manner: A plane leaves each point for the other point every half hour during the day. No reservation is required, and a passenger is always guaranteed a seat on a plane leaving at that time. If a plane should become full with still more passengers to go, the airline brings out another plane immediately. Eastern has been known on a number of occasions to make a flight with only one passenger aboard. This particular approach to providing excellent service has been so successful that many people continue to use it even though it is now not much cheaper than the regular service and it does not provide for any meals or entertainment enroute. It merely shows how appreciative people are about getting good service.

Several years ago a division of Lockheed Aircraft Corporation began using an entirely new approach on processing of accounts payable items. The typical process is having clerical people manually match the following.

1. a copy of the purchase order that was sent to the vendor.
2. a receiving report that was prepared at the receiving dock. This form indicates what and how much was received.
3. a copy of the vendor's invoice which was mailed to the company.

Lockheed recognized that great savings would not occur by merely transferring several of those steps to computers, and decided that radical systems redesign was necessary. So the systems analysts sat down with their internal auditors and began working out an operation that was substantially different from that of the past. Basically the system works as follows: At the time a purchase order is prepared and sent to a vendor, a copy of the order is maintained in computer processable form (magnetic tape). When goods are received, the company's Receiving Department prepares an internal form indicating such and sends it to the computer center, where it is converted into magnetic tape. The computerized purchase order and receiving report are electronically processed by the computer, and when a "match" occurs, the computer prints an accounts payable check to that vendor according to purchase order terms. Lockheed has determined that the vendor's invoice does not play a significant role in the matching and subsequent payment process. Their purchase order indicates what they wanted to buy and the price they were willing to pay. Their receiving report shows what they got. As a result, they then know what they should pay.

Of course, the big problem with the vendor's invoice is that it does not lend itself to computer input easily because of the many styles of vendor's invoices there are. The only practical way to put it into computer processable form would be to manually key-punch cards or key stroke directly to magnetic tape.

A typical reaction by a person who first hears of this procedure is that there are many reasons why it will not work. When many of those issues are pushed aside (most of them are due to a person's reluctance to change), about the only item that could cause a real negative reaction is the possibility of freight charges that might appear on any invoice and would have to be paid. Lockheed has overcome this problem by getting their buyers to negotiate purchase orders with terms of f.o.b. destination. The only other situation in which an invoice might be needed is when there are cash discount terms for paying within a specified time, these terms perhaps not being known at the time the order was issued.

There are realistic reasons why this payment method will not work for everyone that has a computer, but a company should at least take a hard look at it to see if it could be applied.

Many organizations have found a very practical way to save considerable amounts of money in their purchasing and accounts payable operations. It works this way: The company prepares a regular purchase order form which will be sent to a vendor; but they also include an accounts payable check that goes along with the order. The check is signed, dated, and made payable to the vendor, but the amount is not filled in. The vendor physically fills the order, writes in the amount of the check, and then deposits the check in his account as he normally does. The buying company saves many paperwork costs, and the vendor receives his money right away. Your first reaction to this approach might be that a company is exposing itself to too much risk by sending blank checks through the mail. However, those who are using such an operation do provide for various safeguards which are considered to provide reasonable control. Included are the following.

1. Use the system only with very well established suppliers.
2. Limit the amount of such checks to a specific maximum figure such as $1000. Place a time limit upon the check,

such as ninety days, and mark the checks "For Deposit Only."

3. Restrict use of this system to items which tend to have a firm price and those items of a nonspecialized nature which normally would be stocked and available for immediate shipment.

4. Consider each order shipped as being complete with no provision for back orders.

For the typical company in which about 80 percent of all the purchase orders and check writing is required to process only about 20 percent of the total dollar volume of purchases, this system does have some merit and should be considered as a possible approach. Of course, you must realize you can't buy all items this way, and a more normal routine must be followed for many of your other purchases. This means you will have two major streams in your purchasing and accounts payable operations, but if you can apply the alternative described above substantial savings may be realized.

A possible substantial saving in payroll costs and a boost in overall morale may come about by eliminating the hourly payroll approach and the time clock that so often accompanies it. Some companies have found that the hourly payroll basis is an irritant that makes production workers feel they are on a lower level than the people in the office who have usually been functioning on a salary basis. In 1969, the Motorola Corporation did away with the hourly payroll method completely. They put all employees on a salary basis and at the same time provided that each production worker would be allowed five days' sick pay each year. If the person's attendance were perfect he would receive an extra week's pay at the end of the year. Although the president of the company estimated this new procedure might cost about $5 million in the extra week's pay, he was certain that better attendance and increased productivity would more than make up for the additional cost. In order to continue to control people moving in and out of work stations, supervisors were given much greater responsibility to make certain that people were on the job.

A system which many companies have used to considerably cut overall payroll costs is that of depositing employees pay directly to a bank account. Some companies have required employees to use a bank specified by them, but many have extended this to making the deposit in the bank of the employee's choice. Costs that are saved by the company are those of writing individual checks, the cost of reconciling those checks, filing them, and other related costs. Some companies use this approach but actually make out a separate check for each employee and deposit that check in the account. The better approach seems to be to make out only one check for the total of the net pay of all people who are using that particular bank. In this case, if you are a large company whose employees use seventeen different banks, then only seventeen checks would have to be written for this purpose.

The banks are usually happy to go along with this plan because they may end up with accounts they would otherwise not get, and their own paperwork procedures are considerably shortened. In 1965, the U.S. Government authorized all federal agencies to send payroll checks directly to a bank designated by an employee. The overall benefits of this program can be substantial in an organization which has many thousands of employees.

One drawback of this program is resistance on the part of employees to go along with it. But this should be overcome if management will use an educational process to point out the savings to the company, the impact it will have upon profits, and how workers will eventually benefit from it. One company that did not do well selling the procedure to its employees eventually scrapped the method and reverted to the regular way of paying. The breaking point came when a worker said how mad he was because his wife now knew that he really got paid every week.

Controls

In a very general sense, controls are those basic steps which are put in a system to make sure all data gets processed correctly. The concept extends further to try to make all aspects of the system function as they should, which includes more than accuracy. In order to accomplish these goals, a broad line of steps is required.

It may sometimes appear that controls exist only for checking by inside auditors and personnel from the CPA firm. But since management is basically responsible for producing results and protecting the business, they should see that adequate controls are put into effect. Actually, management should be just as interested in them as auditors are.

It must be recognized that you normally get nothing if you spend nothing. Fortunately, the cost of providing reasonable controls will usually be less than the consequences of not having them. However, the control segment of any operation does not have to be another complete system; it can be just those steps carefully designed in (not tacked on) to see that the job gets done properly.

In some transactions, there is a built-in control due to the nature of the situation. In applications such as payroll checks and accounts payable checks, any underpaid amount will usually be noticed by the recipient and brought to your attention for correction. But this type of control is often one-sided, since those same parties can't always be counted on to report overpayments to them. After-the-fact checking procedures may help you to discover the error, but correction steps become costly, you may not be able to collect the difference at all, and undesirable ill feelings may be created by requesting a refund. So it is better to catch errors before they get distributed in output.

Much of the output of systems work, such as statistics as to the profitability of individual items and information relating to the reorder quantities on inventory, is strictly for internal use. Errors made in situations of this type can lead to decisions which might have an adverse effect upon operations. You might continue to handle an item which is no longer profitable, and you may order so little of an item that you are constantly out of stock.

Controls can take many forms, and those chosen should be tailored to the needs of the operation. This is much more effective than adopting eight or ten techniques that may have been heard to work in some other system.

Much of the literature on data processing gives adequate details on controls such as transaction counts, batch totals, and limit checks. It is not my intention to repeat those details here but instead to describe additional systems controls which it is most important to have. Unfortunately, they are often overlooked at the time of estimating systems costs. Then they have to be implemented later when the company is criticized by their auditors for not having them. Or perhaps they have a disaster from which it is difficult to recover, and they then see the changes that must be made.

The controls I want to concentrate on fall into the following categories.

1. *Personnel Organization*. It has previously been emphasized that people are the heart of any system. People at all levels must have some controls placed over them so they perform their

duties in a reasonable fashion. It may appear that the controls mentioned in this section would be used because of an inherent mistrust of people. Although that might be why some classes of controls are necessary, it is not so because of people in general. It is because of that relatively minor percentage of people who try to beat the system to get something that is not rightfully theirs. Since it is not known which people will cause difficulty (they wouldn't have been hired or be retained if you would know who they were), you have to set up the systems so that everyone is under control and would be discouraged from doing something wrong.

One of the ways to do this is to segregate duties so that no one person has complete control over a particular transaction. For instance, the person who makes out checks should not be allowed to sign them and mail them. If that were allowed, there is too much of a temptation for that person to make payments to himself or to others in a manner detrimental to the employer. The person who receives checks through the mail should not be allowed to deposit that money into the bank account and make all accounting entries for it. There is a chance that it will never reach the bank.

A computer programmer should not be allowed to run his computer programs on a continuing basis. Not only would he be able to influence operations too much, but the normal difference between salary grades would mean you would have a too highly paid employee operating the computer.

Another technique to apply is to rotate people among jobs on which control aspects are critical. A person can become much too familiar with a situation if he works on it too long. History shows that much of the internal cheating is done by people who have had the same position for a relatively long time. Not only should job rotation be performed, it is a good idea to do it at unspecified intervals. If a person has been doing something out of line and he knows that in exactly four weeks he is going to be put on another job, he has plenty of advance warning to straighten out his problems. It might appear that job rotation would be a costly practice since some retraining costs will probably be incurred. But the costs of retraining can be small compared to the possible consequences of not doing it. Also, the backup provided by several people to do a job is a worthwhile goal.

It is also a good idea to require employees to take a vacation and have another person perform that function for a while. This not only forces managers to provide a backup for each position, but it also discourages something that might take place if a person knows he will always be there to cover up what he is doing. Many companies have thought it was very nice for an employee not to take a vacation for five years or so. Then they were shocked to find the person was doing something he couldn't afford to have others discover.

2. *Forms Control*. This goes further than the basics of forms design and prenumbered documents described in Chapter 3. It involves procedures to ensure that important input and output documents aren't easily available for indiscriminate use. Documents such as blank checks should be kept in a safe place under the responsibility of a single person. While such a procedure seems to be a natural thing to many, there are still companies who purchase payroll checks with no sequential numbering whatsoever and then leave them in an unprotected area available to many employees. Other documents which must be closely controlled are blank purchase orders and salary increase forms. A signature plate, which is used today by most companies to mechanically sign volumes of checks, must also be guarded so as to prevent unauthorized use.

3. *File Control.* This involves measures for the physical protection of paper forms, punched cards, magnetic tape, and disk files. Protection against disaster would include steps to use fireproof vaults, have smoke and fire detection equipment, keep everything high enough to be above flood waters, deactivate automatic sprinkling systems, and prevent unauthorized entry. Each one is a very specialized situation that requires careful planning. Each will also require some money to be spent.

Recognizing that all such measures taken will not be enough to prevent all disasters, it is necessary to have ways of recovering from disasters should they occur. This can be accomplished by microfilming important documents and preparing duplicate copies of selected tapes and disks. These duplicates can then be stored at remote sites, since a disaster at the main location would not be likely to occur at the other site, too.

The concept of using controls here is similar to that of an insurance policy; a certain amount of money is spent to prevent a loss. The money is considered to be well spent even if nothing ever goes wrong.

4. *Budgetary Controls*. There is the story about the manager of a computer department who used the entire third shift to do service bureau work for outside customers. Each of them paid him through a dummy firm he had established. But it was not through diligence on the part of internal controls which found him out. As so often happens in a case of this type, he eventually made a slip-up which caused him to be exposed.

While there is no sure way to prevent this type of thing from happening, certain measures can be taken to cut its likelihood by a considerable margin. Officials of a company must give more than just a little attention to requests to obtain data processing equipment. They must become familiar enough with work volumes and equipment capability to know if a computer center is spending eight more hours a day than it should. If the company cited above had developed work schedules (or budgets) and then compared actual work to those projections, it would have been easy to see that something was wrong.

In an attempt to do an honest job, the manager of a computer center reported that 8 percent of total processing time in the first month of operation was spent on reruns. His supervisor couldn't understand how that could happen, but he did nothing to try to learn why. The manager of the center was so ridiculed for that item on the report that he never showed it again; every subsequent report buried rerun time in with regular productive work, and he was never questioned again about his productivity.

It is only reasonable that any system manager should be required to prepare reasonable budgets of manpower, equipment, and supply needs. Then certain people in the organization should become familiar enough with that activity to determine if the budget is reasonable. Later, competent people must adequately compare actual spending to the budget. Major differences would have to be reconciled.

5. *Machine Up-Time Controls*. Included here would be procedures to make sure all equipment gets adequate preventive maintenance so that breakdowns are not so likely to occur. Also, when they do occur, you must know of it and be able to act quickly enough to get proper service.

Reruns due to machine problems can be reduced by frequently running data samples and periodically comparing output to what it should be. Keypunch manufacturers recommend that the punching in cards be frequently compared to a standard plate so that adjustments in registration can be made as required.

Contrary to what may appear to be the case, the introduction of automated equipment to systems work has not necessarily imposed more controls than the previous methods did. The only real differences are in the manner in which the controls are exercised.

Inventory Control   (Case problem)

Assume that on the basis of the study outlined in Chapter 6 you have talked with various people and perhaps had a meeting or so with interested managers. You have then learned the following: one of the things the Inventory Control Department does is to prepare a detailed list of every item in inventory at the end of each month. This report is prepared by going through each individual inventory ledger card and multiplying the balance on hand according to the card by the unit cost of the item. The individual extensions are then added together and the total is sent to the Accounting Department to be used as the balance sheet figure for inventory at the end of the month.

Another report which the Inventory Control Department prepares is one which shows inventory turnover as in Figure 7-1. This report is prepared by relating dollar shipments for a month to the average inventory on hand during the period. Shipments are obtained by referring to the appropriate column of each ledger card and adding the quantities shipped during the month and mul-

Figure 7-1. Jones Company inventory turnover report for the month ending October 31, 1971.

Product Group	Beginning Inventory	Ending Inventory	Average Inventory	Shipments	Turnover
1	27,000	29,000	28,000	28,000	12.0
2	141,000	183,000	162,000	81,000	6.0
3	5,500	5,500	5,500	– 0 –	–
4	38,000	62,000	50,000	12,500	3.0
5	– 0 –	4,000	2,000	3,000	18.0
6	102,000	106,000	104,000	38,000	4.4
Totals	313,500	389,500	351,500	162,500	5.5

tiplying by the unit cost. The average inventory for the month is arrived at by adding the balance at the beginning of the month to the balance at the end of the month and dividing by 2.

The average is then divided into total shipments. Since a turnover is more meaningful when stated on an annual basis, the resulting figure is multiplied by 12. In order to eliminate the need for showing this precise detail on each and every item, the company has always pulled certain categories of items together and shown an average turnover for that particular group.

Another general report which Inventory Control prepares is a basic summary of transactions processed during the period covered, the number of requisitions sent to the Purchasing Department, the number of vendor shipments of goods received, and the number of shipments made to customers as indicated by the shipments column on the inventory ledger cards.

Analysis of a sample of inventory ledger cards shows that all of the transactions recorded on the ledger card are made according to the date at which the posting is actually made. In most cases, the date does not correspond to the date when the transaction itself occurred. For instance, any entries made in the reserved column reflect the date on which the entry is made in the column and not necessarily on the date on which the customer order was received or the date on which the order entry section prepared the internal form that is used to eventually make the shipment. Likewise, receipts as posted reflect the day on which the receipt was posted and not necessarily the date on which the material was physically received in the Receiving Department. Because of this time lag, inventory values and all statistics furnished have discrepancies. While it is true that these figures tend to balance out over a period of time, the reports rendered in any one month are not representative of what has happened.

Further investigation reveals that on each Thursday afternoon analysts in the Inventory Control Department review all inventory ledger cards. When any item has a balance in the available column which is lower than the reorder point shown in the upper right hand corner of the ledger card, the analyst sends a requisition to the Purchasing Department. So long as the amount shown as the reorder quantity would be great enough to bring the available column up to a point in excess of the reorder point, only one order unit is placed for the amount of the reorder quantity. In the event, one order unit is not a sufficient amount, then the unit quantity is raised by those multiples necessary to bring the available column back in excess of the reorder point. If available column shows 5, the reorder point is 20, and the reorder quantity is 8, the analyst would order 2 units of the reorder quantity, or 16, to raise the new available to a point higher than the reorder point. This latter situation would arise only when many large orders were received in the same week, since reorder points and reorder quantities are developed on the basis of average usage. This particular point is illustrated on the line for February 4, Figure 6-5, where 4 order units were required to raise the available above the minimum of 25.

Additional checking with the people in the Purchasing and Traffic Departments reveals that because requisitions are received by Purchasing only once a week, the department has a peak load of requisitions to process so that purchase orders can be placed with vendors as quickly as possible. Due to the great rush of activity in that department it is very difficult to have sufficient time to do any analysis which may have a very desirable effect. As an example of something that Purchasing would like to do, a more careful study of the total number of items being ordered from any particular vendor may result in the possibility of taking advantage of quantity discounts on the value of the total order placed. Perhaps the Traffic Department may be able to allocate orders in such a way that a full truck load or full railroad car load could be ordered from a particular point to take advantage of considerably lower freight rates.

Because of the problem cited in the previous paragraph, incoming shipments also tend to bunch up in the Receiving Department and there are perhaps two or three days of the week when that area is very busy and on the other days of the week activity tends to slacken off. Peaks and valleys in Inventory Control cause peaks and valleys in the purchasing cycle which cause the same thing in the volume of shipments as it flows into the Receiving Department.

Another interesting item is found that creates a considerable discrepancy in inventory turnover figures. The shipments column of the ledger card includes any shipments made to the company's subsidiary in Buffalo. When Buffalo requests a shipment of material from the company, it is done in a manner whereby the material is taken from stock or it is purchased and received directly at the home office; then on the same day it is sent to their Buffalo station. As a result the shipments figure for the inventory item reflects that quantity as a shipment in the same manner as it would for any other shipment made to a regular customer. What this does is seriously inflate the shipment figure resulting in a turnover figure considerably higher than it truly is. Inventory analysts have always been praised for the high turnovers which their reports show, but it is now relatively clear that they should not be getting any credit for a high turnover when their method of calculation includes something that is completely beyond their control. Figure 7-2 illustrates this point.

Figure 7-2.   Effect upon turnover figures due to Buffalo subsidiary operation.

	Beginning Inventory	Ending Inventory	Average Inventory	Shipments	Turnover
Figures from 10/31/71 report	313,500	389,500	351,500	162,500	5.5
Less: Shipment to Buffalo	–	–	–	42,500	–
Results for material contolled by home office	313,500	389,500	351,500	120,000	4.1

Although the discussion above has been described from the standpoint of a system using a manual system revolving about ledger cards, the same operating problems could have occurred in a computer operation using cards, tape, or a disk.

New System Design

Because the company has apparently felt that not everything in the inventory control operation is moving ahead as well as it might, the Systems Department has been called in to make the study that is now under consideration. As previously stated one reason for making a study is the knowledge that operating results are not compatible with the plan that was established.

As you have observed the activities of this hypothetical company relating to inventory control, you should now realize that inventory control truly involves more than just those people who sit in a room that has the sign *Inventory Control* on the door. You recognize that many departments are involved, and certainly most of the employees of the company are affected by the method by which inventory is controlled.

As a first observation you recognize that both the Purchasing and Traffic Departments feel they could do a much better job and perhaps lower the cost of purchased items as well as related transportation costs if they could have a little more time to spend on planning for maximum results. You, therefore, determine that if purchasing received requisitions from inventory control on a daily basis instead of on a weekly basis, they would be able to do the following.

1. They could plan their activities more carefully.
2. Peaks and valleys in the operation could be cut back.
3. Purchase orders could be processed so that goods would flow into the company on a regular basis. Goods might arrive on an average of two or three days sooner if purchase orders went out on an average of two or three days sooner, and reorder points could be adjusted accordingly.

In keeping with the turnover objectives set forth by top management discussed in Chapter 2, we would now determine that a more accurate means must be used to calculate inventory turnover. It is going to be much more desirable if turnover will include only those transactions over which the Inventory Control Department has direct control. Therefore, it would probably be your feeling that any shipments to the Buffalo subsidiary be removed from the calculations used in arriving at the inventory turnover. Because of the manner in which records are presently kept, it is relatively easy to go back over a brief period of history and recalculate the turnover figures. As a result of doing so on a sample basis, you would find that realistic turnover is considerably lower than what previous reports have shown, and in fact turnover has steadily declined over the past two years. For an illustration of this point, refer to Figure 7-3. What this shows is that management has been

Figure 7-3. Real turnover after giving effect to shipments made to Buffalo subsidiary.

Report for month of	Average Inventory	Shipments	Reported Turnover	Shipments to Buffalo	"Real" Shipments	"Real" Turnover
June, 1970	302,000	136,000	5.4	16,000	120,000	4.8
Dec., 1970	331,000	152,000	5.5	28,000	124,000	4.5
May, 1971	346,000	158,000	5.5	33,000	125,000	4.3
Oct., 1971	351,500	162,500	5.5	42,500	120,000	4.1

lead to believe a condition which obviously is not the case. Since reports showed a steady turnover, management had no reason to take a closer look at what was happening.

However, in any of your discussion and writing you must be careful so that the people in the Inventory Control Department are not criticized because of what you have found. There is nothing to be gained by focusing attention on errors that have been made, so long as the new system does not include ways for them to continue. Also, there must be some assurance that something will be done with turnover reports other than just putting them in a filing cabinet.

As you continue your investigation, you check within the company to see what possible use is made of the various statistics furnished by the Inventory Control Department. A specific area concerns the report showing the number of pieces of paper processed. You may find that nobody really does anything with the reports. In every case, you find the recipient of the report merely files the document in his filing cabinet and never refers to it again. In fact, your penetrating questions reveal that the report was requested about three years ago when the manager of the Inventory Control Department was attempting to hire a new person for his area. He used the report to show the increase in transactions processed over the previous year and that one person could only process so many in a day. Thus he was able to obtain the extra clerk requested. Since that particular type of request has not been made again within the past three years, there has been no continuing need for the report. Your recommendations regarding this particular item would no doubt be to discontinue it because of its lack of value. Realistic steps must be taken to make sure the Inventory Control Manager himself has no valid use for the data.

There is some question regarding the method by which the calculation is made to arrive at the monthly inventory figure for balance sheet purposes. You discover that so much time is spent at the end of the month in arriving at the figure that either overtime must be consumed or the regular work of the Inventory Control Department is held up until the report can be prepared. In talking with internal and external auditors and also members of the Accounting Department, you find that all of them would accept an inventory figure which would be arrived at on an estimated basis. However, they all agree that the inventory figure appearing on the company's annual report balance sheet should be arrived at as a result of having taken a physical inventory rather than using an estimated figure. You are willing to recommend the information be provided in this way and thus save all that detail work which is now being performed each month. At least the company will be able to reflect the savings of this detail work in eleven of the twelve months. In order to back up your feelings on this point you may be able to apply the proposed process to some of the previous balance sheets and show what the net difference would have been between detail calculation and estimate.

You may logically conclude that a lot of effort can be saved by this proposal if the company functions on a manual basis; but it may be pointed out that only one computer run is necessary if the data is in machine-processable form. But an additional computer run at the end of the month may be just what you can't afford because of other pressing jobs. Even though the annual printing of employees W-2 forms represents a relatively simple computer run, many companies are hard pressed to get the forms in the mail by the January 31 deadline.

Another major aspect to consider is that there may be a considerable waste of time due to maintaining both available and on-hand figures on the inventory ledger records. Study shows that often material is physically on hand and shipment could be made at least several days faster if shipping forms could be sent to the Shipping Department on the same day in which the item was known to be on hand. The overall effect of a proposed change in this procedure would eliminate one complete posting and one complete passage of inventory transactions against the file. A company that was still using this method in the 1960's had adopted the double posting during World War II. With material shortages and occasional delays in freight movement, the company had used the procedures to make sure that all shipments were allocated to customers according to management wishes. The reason for the procedure had expired many years ago, but the company had never taken the steps to revert to a more meaningful process.

Another major benefit could occur in this overall program if it would be possible to have someone work more closely with the reorder points and reorder quantities. If a sufficient effort could be spent on this particular aspect of the operation, then the company may be able to get by with a considerably lower overall value of inventory. For instance, if the individual reorder points were originally set by someone who had the belief that the company should never run out of any inventory item, then it is time to go back through and review to see how realistic these points appear at the present time. For instance, management would certainly be interested in knowing how much inventory could be lowered if reorder points were established in such a way that on the average an item would run out of inventory only once a year, twice a year, etc. Since most of the other activities of the company relate to some form of risk taking, Inventory Control should at least look at the same approach and see what benefit it might have to them.

Questions

1. What is a transaction? Give five common business examples.

2. What is the basic difference between off line and on line? batch and real time?

3. About how fast must transactions be processed in order to be considered real time?

4. What generally must be considered in order to determine the best way to record transactions?

5. On what phase of operating a system is the most time spent?

6. It has been mentioned in the text that you should not spend all your time trying to handle things as they are but to sometimes try to change operating conditions. What can be done about the great peaks and valleys in activity in the following situations?
   a. A state having all drivers licenses expire on the same day.
   b. A telephone company billing everyone on the last day of the month.
   c. A city collecting all real estate taxes in a two-week period.

7. Why didn't Lockheed Aircraft just mechanize their accounts payable operation as it was? (See text.)

8. Make a list of reasons why inventory turnover might be low. (Note: Is it fair to say there is not much demand for your products?)

9. What is the proper quantity to order if available is -325, the reorder point is 50, and the reorder quantity is 150?

10. Match the situation on the left to the type of processing system most likely required from those on the right.

   a. Inventory control   1. Real time—random
   b. Payroll   2. Batch—random
   c. Savings accounts   3. Batch—sequential

11. A large company has its major markets in and around Chicago, Los Angeles, and New York. It has a billing center in each city. The present system calls for all customers to send their checks to Chicago, where they are deposited. But there is an average of three days after the customer mails his check until the company can make its deposit in its Chicago bank. Briefly describe a new system which would eliminate most of this "float." Carefully consider how much data must be moved to Chicago each day, as that is where the money must be to satisfy the company's needs.

12. The following ad appeared in a leading financial paper: "Put your money in California Savings & Loan. We compute interest on a daily basis. $1,000 deposited here at our 4.9% rate returns you $50.22 in a year. $600,000,000 deposits strong to serve you. Join 500,000 other happy depositors."

   a. In the broadest sense, what problem was the bank trying to solve?
   b. Most likely the depositors accounts would be stored in what form? Why?
   c. When would interest calculations be made? Why?
   d. How would input enter the system?
   e. Even though interest was not to be paid on accounts of less than $5, what should they do to make sure they don't have serious rounding problems on small accounts?
   f. Briefly summarize the nature of costs of a system like this.
   g. After the first year of operation under such a plan, how could they determine whether the system was paying off?

Other Readings

Data Communications in Business: An Introduction. New York: American Telephone & Telegraph Company, 1965. Chapter V gives an excellent approach that can be used as a guide for designing any system or solving any problem.

Lott, Richard W., *Basic Data Processing*, 2nd ed. Englewood Cliffs, N.J.: Prentice-Hall, Inc., 1971. Chapter 14 illustrates common methods used to control data processing operations.

Nadler, Gerald, *Work Systems Design: The Ideals Concept*. Homewood, Ill.: Richard D. Irwin, Inc., 1967. Full of very helpful ideas on designing better systems.

CHAPTER 8

Benefit and Cost Analysis

One of the very difficult things for the beginning analyst to realize is that quite likely a new system may not cost less to operate but more than the old system. The point is that you are not likely to get something without spending money, or as the saying goes you have to "spend money to make money." Typically, a company has had to spend money in order to make a better product or to provide a better service, and they usually have to do the same to design and operate a better system. For instance, the proposed solutions to most of the problems of the U.S. Post Office would require billions of dollars to put into effect.

Of course, this does not mean that you won't strive to reduce costs if you can, but you should not expect as a normal consequence that costs will automatically decline. Lower systems costs tend to come about because you have eliminated procedures, not because you are performing all the old steps in a different way. So many new systems over the years have been installed after they have been justified on the basis that new system costs would be lower. In many cases, the results don't turn out that way, and then management becomes very disenchanted with the people who have participated in systems analysis and design. There are reasons for the unplanned operating results which often occur.

1. Inadequate knowledge of the things to be done causes unrealistic estimates to be made. Estimated times and costs can so easily be too minimal.
2. Analysts very often think management wants to hear about savings even though analysts themselves may not believe there are going to be any. This point took a strange twist in one company; the president asked an analyst to justify something of which he was very fond. The study showed it would be a losing proposition and was so reported to the president. The president clearly pointed out that the report would have to be done over showing some savings.
3. The analyst may have fallen in love with the system so much that he is overselling it. Or an individual may have so much personal interest in a system that he has enough power to see that it is put in.

With nothing in view to indicate that inflation in employee costs has stopped, with employee costs continuing to represent the major cost of most operations, and with the volume of business continuing to rise, it is only natural to expect most systems costs to continue to rise. Despite the fact that systems have been with us a long time and therefore probably have been worked on enough to make them efficient and that highly automated processes have been generally available since the late 1950's, you probably can't find many organizations that have reduced costs over the past few years. Recognize that this statement refers to total systems costs and not just to costs of specific procedures. Due to the use of much mark sensing and a very sophisticated scanner, a company may have greatly reduced keypunching costs, but other operating costs may have increased tremendously.

Various studies have been made to compare the unit cost of a computer calculation today to one of ten or fifteen years ago. A typical study may show that the cost today is only one-thousandth of what it was previously. But no computer department today has costs of only one-thousandth of what it had fifteen years ago. Because of the nature of the differences in programming, today's computer must perform many more calculations just to execute the program. The computer today is also running more programs, and it probably has more peripheral devices attached to it, each of which may require instructions to properly handle. For these and other reasons, you are not likely to find a computer department which has enjoyed a downward cost trend.

It may, therefore, be a sounder approach to be very realistic on timing and cost estimates, and question any systems costs that are projected to decline. This is preferable to being too optimistic. A good friend of mine has always followed a policy of doubling all computer timing estimates and adding about 25 percent to cost estimates. His reasoning is you can't plan for all the things that will have to be done or those that will go wrong and need rerun. The equipment can be run overtime one whole extra shift to make up for the time they need so it is not all costs that are doubled. Just a few additional machine costs are incurred, and only a certain number of additional people would need to be hired or work overtime, so the staff has to be increased by only a portion, not 100 percent.

Costs often seem to be very unusual. For instance, some companies have been able to considerably reduce the cost of heating their buildings after they have installed lights that provide sufficient heat to warm the interior. Such buildings have another heating system installed so the buildings can be kept warm during the time period when the lights are not on. Another interesting cost aspect is companies with extensive lighting fixtures have been able to save money by replacing all the bulbs at the same time. Instead of following the usual practice of having a workman go through the steps of replacing a single light when it has burned out, the major problem is determining from the bulb manufacturer or other sources what the expected life of a bulb is. Then the company replaces all bulbs as soon as the group has been used that many hours. It may seem like a waste of money to throw all those bulbs away at one time when obviously most of them are capable of at least a little more service. But it is the cost of the workman replacing bulbs on a nonscheduled basis that represents a significant cost, the elimination of which makes the whole program worthwhile. Despite well-documented articles showing the value of such a practice, there are many people too skeptical to put this one into effect.

Proper cost analysis is something which must be going on all the time a systems effort is underway, not just during a single phase. Certainly a company has some "ball park" figure in mind that they are willing to spend before a complete system redesign gets underway. If such a figure is available, analysts can constantly be reviewing to see how their work is progressing. They must be able to properly consult with management at all times and be in a position to have the whole project scrapped as soon as it would appear that total costs are going to be too great or that adequate benefits can't be obtained. One group of analysts, when considering the purchase of a computer, made a very rough sketch of the costs of computer operation before they ever designed the system in detail or even went out to vendors to prescribe the type of equipment they would need. Fortunately for the analysts, it was learned very early that a large figure was completely out of line with what was needed and they were able to immediately stop the process before they had spent a lot of time for which nothing would be gained. This is why it is desirable to perform some cost analysis along the way so that too much effort is not expended in an area where there will not be an adequate return.

Benefits to Be Gained

You can logically ask why you would ever want to install a new system if the operating costs are not going to decline. The following summary of a hypothetical situation should provide a sound answer to that question.

	Before	After	
		Increase Sales	Decrease Other Costs
Sales or total revenue	$100,000	$102,000	$100,000
Systems costs	7,000	7,150	7,150
All other costs	82,000	83,500	81,300
Total costs	89,000	90,650	88,450
Net income before taxes	11,000	11,350	11,550

Note that systems costs, the costs of collecting data and generating information, are a relatively small portion of the total. Savings, or net income, can be increased by increasing sales or by cutting those other costs. Even if a company can't increase sales, there are probably many areas in which to make substantial improvements in those other costs.

Of primary importance regarding any systems work is the overall effect upon the services you are rendering and, hence, upon net income. This is the only area in which there is a true payoff. Consider the following instances where "systems" costs went up considerably but there was substantial return as shown.

"System" Change	Payoff
Airlines replaced $5 million 707's and DC-8's with $22 million 747's	Much better passenger/employee ratio and lower plane costs per passenger mile
Hershey Foods Corporation began spending millions per year on advertising.	Made introduction of new products much more successful in a shorter time period
An automobile manufacturer spent $500,000 per year checking out automatic transmissions	Warranty costs were substantially cut because defective transmissions were caught before installed in cars
J. C. Penney Company installed a multimillion dollar computer system to handle charge accounts	Brought the company to level competitive with other chains and helped increase profits by increasing sales and producing service charge income
Internal Revenue Service installed eight mammoth computer centers to handle income tax processing	Much better collection procedures underway, and fewer people able to avoid taxes

It may also be that having a new system will result in spending considerably more in the first year or so due to a combination of high design and implementation costs. Then after the system is under operation for a while, costs may tend to drop a little as design costs disappear and peak operating efficiency is hit. Management must realize this cost pattern is most likely to occur, and they should not be led to believe you can install something and have huge savings immediately.

You must develop a general understanding of total costs and unit costs. On the income statement, where it counts, the emphasis is on total costs. But if you can do something so that the value of each transaction goes up $1.50, if the unit costs of processing each goes up only $1, and if you get thousands of them each year, that is a significant contribution to total profit.

On the other hand, you can go only so far in spending increasing amounts on systems. In one small operation, it was determined that the company would have to spend $100,000, $150,000 and $225,000, respectively, as additional costs in the next three years in order that substantial benefits could begin in about five years. Management decided there was absolutely no way for the firm to finance such a venture at the present time regardless of the eventual benefits, and the recommendation was not accepted.

The opposite situation occurred in one of the regional scheduled airlines. A few months after they had installed an intricate on line real time seat reservation system, they said they had made no detailed comparison of costs to those of the old system. They had gone ahead strictly because they had to in order to regain a competitive edge that had been slipping. Potential passengers were known to first contact competing airlines with real time systems because they knew they received better reservation service there. Management of the airline recognized that other airlines of the same size had done so profitably, and it was something they had to do to stay in business.

Inventory Control

It should be clear that your precise objective may not be to develop a system that is less costly than the present one, but knowing the cost of it will be of some value to you. There are various ways of obtaining the cost of the present system, several of which would provide estimated but quick and usable answers.

1. Look in the general ledger for the cost in the accounts identified by the appropriate department name. Recognize that

there will not always be a department shown for the system you want, and any department that is shown is not going to present all the costs for the application. We have already seen that many departments other than Inventory Control incur costs regarding that application. In analyzing figures arrived at by this method, you may want to carefully look into the method by which costs of other departments are allocated to this department. For instance, if costs are allocated from an area like the Personnel Department, they could be eliminated because they are not relevant. In the following example, $55,700 ($61,700 — $6,000) would be the relevant figure.

Account Name	Amount
Direct payroll	$ 45,000
Fringe benefits	8,000
Travel	1,500
Supplies	1,200
Service department allocation	6,000
	$ 61,700

2. Prepare a flow chart of the total operation and cost out each step of each department that contributes to Inventory Control. This could be done in a general sense or in a very detailed way.

3. Send a questionnaire to all people involved and have each of them show the percentage of time he devotes to this application. Then put a dollar value upon all the time. Some fallacies of this approach are that you may tend to overlook some people who are involved unless you send the questionnaire to everybody or some applicable people will not bother to send the form back. Also, people might have a misconception about the amount of time they spend at various tasks. For instance, if the same people were later asked to fill out a comparable form on all their other applications, there is a great chance that the total time they report would exceed 100 percent.

4. Determine the total purchase cost of all business forms used with respect to the application and multiply by a factor of 20. You will recall a previous statement that many systems are believed to cost approximately 20 times as much as the cost of the forms used.

5. Try to find out from your industry association what the average relationship of cost to sales is and then multiply that rate times the appropriate figure for your organization. If the industry rate on Accounting Department costs is 3 percent of sales, just

multiply 3 percent times your sales figure. Actually, this method which uses the norm for your industry will more nearly approximate the amount that might be appropriate as opposed to determining the amount you are actually spending now.

Of course you must seriously consider the value that attaches to knowing what the specific cost of a system is. Perhaps you should be more concerned that profits under the new system will be higher than they are under the present one and therefore you may not be too concerned about the absolute value of costs at the present time. If there are fewer people in the new system or if certain steps are eliminated, you can be fairly sure that it costs less even though you may never have known specific values of each.

You will be seriously shortchanging yourself and your company if your analysis of present costs stops at the point where you have added up to get a total cost for inventory control procedures. You must spend considerable time determining what the effect of this operation is on the costs of various other segments of the business. For instance, you must be able to determine what the net effect on increased capital needs or borrowing is as a result of a particular level of inventory. Obviously, if your inventory control procedures are poor and you have more money tied up in inventory than you could otherwise get by with, interest costs are going to be higher than they would have to be if you had a better method of controlling inventory. If top management has a false impression because of fictitiously high turnover figures, perhaps they never had reason to believe that inventory may be out of line. If inventory is $100,000 higher than it should be, excessive interest costs alone could amount from $8000 to $15,000 annually.

Other costs, related to obtaining and holding inventory, can get out of line due to an improper system.

1. If the manner in which the reorder cycle is handled and purchase orders eventually prepared results in sending a great number of single-item orders to various vendors, something must be altered. Suppose a company determines that it costs $8 to process its average purchase order. They could save $4000 if they could reduce the number of purchase orders per year from 3000 to 2000, even if the unit cost went up to $10.

    $$[(3000 \times \$8) - (2000 \times \$10) = \$4,000]$$

2. If there is no true effort to consolidate shipments to obtain purchasing economies, the system needs changes.

3. Is an effort made to consolidate shipments to obtain quantity discounts on freight rates? Suppose two 30,000 pound shipments each carry a freight rate of $1.40 per hundred pounds. The company could save $240 if they consolidate them into one shipment at $1 per hundred.

$$[(2 \times \$1.40 \times 300) - (\$1 \times 600) = \$240]$$

Other possible costs which can get considerably out of line are those related to warehousing, taxes, storage, theft, obsolescence, and insurance. By having too high an inventory, each of those costs is likely to be somewhat higher than it need to be.

A cost that can be much too high because of poor inventory control procedures is the cost that is incurred in warehousing activities. If great peaks and valleys occur in the shipping operations because of the procedures described in Chapter 7, it is likely that labor costs in the warehouse are higher than they should be. Many companies have found their warehouse personnel idle at one time of the week and yet they have to pay them overtime several days later.

A very obvious cost that can become a considerable burden is one caused by obsolescence. This is the loss incurred as a result of throwing things away after they are no longer salable at their normal value or the substantial reduction in selling price to recover only scrap value. In 1962, I visited a very large warehouse full of parts for steam locomotives. That was at a time when railroads had virtually reached 100 percent dieselization. (A few lines have segments operating on electric power.) The owner must have suffered a several million dollar loss as the result of having too much of a product for which there has been no further market in the United States.

So you can see that the approach taken here in analyzing present costs of inventory control goes far beyond the costs of the Inventory Control Department. An attempt must be made to relate very specifically to all the other operations of the company that are directly or indirectly involved. Obviously, some of those costs outside of the department can be cut considerably by merely improving what the system does without respect to the physical means by which information gets processed. Computerization of clerical procedures in the inventory control process may raise total costs in the Inventory Control Department, but the costs that are being incurred outside the department may be favorably affected regardless of the tools employed.

Notice that the approach taken here is not one of talking about intangible things such as improved customer service or the value of getting reports more quickly. Instead, the analysis relates to very specific items of cost and value which can and must be reasonably measured. Too many systems are justified on the belief that all of these intangible items will somehow work out to everyone's satisfaction shortly after installation. But in our case we have provided for a sound method of measuring what systems improvement will accomplish for the organization without relying upon things that can't be measured.

In the event that substantially improved inventory control procedures are successful in reducing the amount of money tied up in inventory and at least as good or perhaps better customer service is being provided compared to the previous service, you may wonder what the real benefit of that will be to the company. There are a number of good uses to which the company may be able to put the money.

1. If the company is presently borrowing money, they may be able to repay these loans. If the need to borrow is imminent, they may be able to eliminate future borrowing. Even if the amount made available should fluctuate on a day-to-day basis, there is a ready investment market to be utilized even for only a night.
2. The company may have other projects in mind for which they need money, and the source for such funds may be what is freed from inventory.
3. Perhaps the company has been overcapitalized. The freeing up of these funds will permit the company to buy back some of its own stock and thus permit a greater rate of return on the outstanding stock. If you own 5 percent of a company and the inventory reduction is sufficient to reduce the outstanding stock by 10 percent, you would then own about 5.5 percent. [(Original 100 shares minus 10 shares) divided into your 5 shares equals 5.5 percent]

Notice that the lower investment in inventory will have great monetary benefit to owners even though total costs of operation may not change at all. However, based upon all the other costs mentioned in this chapter which may be reduced as the result of a better system, there is a great possibility that total profits may also increase and thus provide benefits from two sources.

Suppose that you own 5 percent of a company that has 100 shares of stock and which normally earns $20,000 a year. The per share earnings would be $200. If systems changes increased profits by $2000 and also enabled the company to retire 10 shares of stock, the per share earnings would jump by 12 percent to $244 ($22,000 ÷ 90). Your own share of profits would climb from $1000 (5% × $20,000) to $1210 (5.5% × $22,000).

In attempting to calculate inventory turnover values on a reasonable basis, it may be possible to do a great deal with sampling techniques. Instead of going through the records and calculating the turnover of every item, it may be possible to calculate turnover on every tenth item or perhaps every twentieth and so forth. If it can be determined that a reasonable sample can be worked out, this approach may be very valuable to satisfy management needs. Of course, the financial people can always work out an overall inventory turnover figure from their financial statements, but you must remember any number obtained in that manner is an average. It will not begin to suggest what some of the highs and lows actually are for the multitude of individual items in the company. For this reason, it is necessary to see the calculation made on every item at some point in time.

With respect to the theories advanced and estimates used in this chapter, it is absolutely necessary that operating people review all of them to make sure that they are in reasonable agreement with the logic you have used. In fact, you will be on much safer ground all around if you can get the operating people to make many of the estimates for you based upon their substantial knowledge of the data. This should also be easier after a responsible discussion with you in which you show how the new system may operate. If your study of the system is well handled, you may even be presented with evidence from the past that indicates benefits could have been attained if something different had been done, but because of various reasons had never been put into effect.

In summary, operating people should be greatly involved in a study of the costs and the benefits.

Questions

1. Unless certain processes are eliminated, what is the likelihood that the cost of operating a new system will be less than that of the old? How then, can you justify any system changes?

2. As computers have been built to operate faster, have the costs of most computer centers declined? Why?

3. Why might a company decide not to install a new system that practically guarantees a substantial return? Why might a company decide to install a new system that they know will only break even?

4. Do you believe it is a good idea to completely overlook intangible benefits when attempting to justify a system? Why?

5. How can a system be of any advantage if it does not increase net income? (See 3 above.)

6. Assume that a company has studied several different ways of distributing pay checks to workers and that having the supervisor give them out at the work station is the most expensive of the alternatives. Why might a company pick this method?

7. Why hasn't the U.S. Post Office made first class mail rates vary according to distance? Could they do so?

8. Should an electric company bill its customers every month, every two months, or how often? Make a brief outline as to a good way to arrive at the best billing interval.

9. A company uses a 50,000-name mailing list every three months. It takes twelve hours of computer time to process the file each time. A computer hour is costed at $75. Total supplies costs amount to $5000 each quarter, and postage totals $3000. The company estimates that 5 percent of the names in the file are duplicates. If it took ten hours of computer time to print out the file contents for a clerical review and one minute of clerical time to locate and remove each duplicate, can they afford to purge the file if they expect its life to be three years? (Assume there are no additions, and disregard deletions other than those due to duplication.)

10. A football team normally played its home games in its own 35,000 seat stadium. A championship game was coming up, and it appeared that the number of people double the seating capacity would want to see the game. Management of the team requested the renting of the local university stadium which had adequate capacity. But they were turned down because of a state law that did not permit use of a nonprofit arena by a for-profit business. What reasonable thing might have been done to satisfy all concerned?

CHAPTER 9

Selecting the Best System

Any system that is built too rigidly will not serve well because it is not able to meet the various and changing demands which you can be sure are going to occur. The purpose of this chapter is to present some ideas which you might use so that the system which is finally selected will provide the necessary outputs with proper allowance for flexibility.

Providing for Alternatives

You as the analyst will not do an effective job if you accept the first idea that comes to your mind and then insist upon adopting it as the way that job will be done. You must resist any urge to do this but instead must spend the time to uncover alternative, acceptable ways. It is only through this approach that you can eventually weigh one method against others and arrive at the best one.

Generally speaking, systems analysis and design involves processes to accomplish things in a better way, but since there are no doubt varying degrees of better ways, it makes sense to take a little more time and set it up the best possible way. This doesn't mean that you will strive to reach a solution which will carry the application for the next hundred years. Five years from now you could be again working on the same application, and the conclusion you reach at that future date may be drastically different from the one that would apply now. Conditions change often enough to require changing solutions. Examples of changing conditions are:

1. volume may climb significantly.
2. management changes may have altered information requirements.
3. owners may require greater returns.
4. technical improvements may have made other solutions feasible.

You must develop the habit of finding out how many practical ways there are of satisfying the requirements of each and every function. Obviously the person who has painted himself into the corner of a room or sawed off the tree limb he was resting on could better have followed such advice. Perhaps those two examples have always been used in a humorous sense, but there are numerous cases in business where details of the system were not that well thought out, and similar chaos resulted on processing of certain types of transactions or perhaps on every transaction that occurred.

As a computer operator was about to run the program to print pay checks, he pulled his own payroll card from the file. (He was just curious as to what was punched in the card.) After looking at the card he inserted it at the wrong place in the deck, which changed the check run. It turned out that twenty payroll checks were not written at all. Because of fairly loose controls, no one knew the checks were not printed until it was time to hand them out. There was no known way to write the twenty checks on the computer without doing the entire batch over. Finally, the managers of the Computer and Payroll departments decided that the only way they could rectify the situation was to make checks out to the twenty individuals by hand. They no longer had the time nor the means to manually calculate the exact salary for each, so the checks were made out for an amount larger than usual. When the exact pay would be calculated by the computer the following week, the excess payment could be taken out as another deduction. If the twenty pay checks had been made up in amounts too small, there was no way in the computer program to add the deficiency to next week's pay.

One business had a very unusual thing happen in its cycle billing operation. (Cycle billing is accomplished by making out invoices to customers only on specific days; perhaps the accounts starting with the letter A are billed on the first of the month, the B's on the second, etc.) When they heard that one of their customers was about to become bankrupt, they had no way of immediately generating an invoice to him because they had so rigidly designed the cycle phase.

On the other hand, the following example shows a very responsive reaction. In September 1968, $250,000 worth of Social Security checks took ten days to travel from the San Francisco Treasury Department office to Sacramento, 90 miles away. No reason was discovered as to why this particular shipment took so long. As soon as the checks were known to be missing, the Treasury Department prepared substitute checks and had them ready to distribute when the stray mail bag was found in a post office.

Any presentation you eventually make to get approval of your proposals will be much better received by your audience or readers if you can show some of the various solutions you have considered. By giving reasons for which you have accepted or rejected them due to relative advantages and disadvantages, you can show how objective you have been. Operating people will feel much more like cooperating with someone who shows that he knows what he is doing. Remember that their present system is working, even if it may be working poorly, and you probably need them more than they need you.

By having followed this broad-minded approach you are also in a better position to assess ideas and recommendations of others. If you haven't even considered a certain point, it may take a great while before you learn enough about it to reply in a knowledgeable sense. If you are at a meeting where relevant topics are being discussed, you can hardly keep saying "wait until I get a chance to check that one out." You are in a much better bargaining position with management and operating people if you can display your knowledge of what is going on.

Do the Alternatives Really Do the Job?

One of the basic reasons for making a complete systems study is apparent weakness in the present system, and someone has felt that major changes are necessary to bring it up to the level at which it should be. If you are at a stage where you are ready to make the decision as to final selection on the proposed system and if it looks like the proposed one is only a warmed-over version of the present one, there is nothing that can be done about the money that has been spent to date. There is no way to turn back the clock; the money that has been spent is an expense regardless of whether you adopt the new system or not.

Suppose your goal is owning and operating a car for the smallest possible cost per mile. Suppose you had spent $6000 for the car you have now and it would require an additional expense of $2000 to get it in shape to provide service at a projected cost of 12 cents per mile. Suppose, on the other hand, you could go out and buy another car for $2000 that would provide service at 11 cents per mile. Based on your objective, it would not make any sense to proceed with the original car; the amount of money that has already been spent ($6000) is irrelevant and does nothing to make that situation any better. Notice that an intangible such as one's own pride may cause a person to make the wrong decision on a point like this.

The illustration on the car can be applied to systems design costs, too. There is no such thing as having spent so much already that you can't turn back if the results obtained by proceeding with the plan aren't adequate.

The important point that must be stressed here is that the business still has a chance to save itself all the problems of conversion by not putting the new system in if it isn't going to do what was needed. Typically there are so many disruptions involved as you take the steps needed to make a major systems change. If it appears the results of the new one would not be that much better, you may want to stop further work on it before you spend more time and money fruitlessly. In this connection it is desirable to have negotiated all contracts so that you will be able to back out as late as possible without penalty.

On the other hand, there should be no changes made merely for change sake. There should be some sound, logical basis for alteration other than a general justification that it "was time for a change." An area in which this principle is consistently violated is professional sports; coaches and managers are fired and new ones hired just to give the team a new look. In most cases, the manager hasn't lost his ability to run the team, and top management just makes the change that is easiest when the team is not winning. Normally the reason for not winning is the lack of the right players, and since the right players can't be easily obtained, the easy step of getting rid of the manager is followed. If you would compare the immediate results under any new manager, you'll find them not much better than under the previous one unless new players are obtained or enough time has passed for the players to mature. Neither of those causes may bear any relation to the work of a new manager.

Don't Design for Today

A certain approach must be followed so that you don't design the system for the past, present, or just tomorrow. It must be set up well enough to provide for the reasonable future. Quite often the whole series of analysis and design steps from beginning to installation may be measured in calendar years. This means you aren't trying to satisfy needs as they exist at the beginning of your study but as they will exist at the time it is complete and for two to five years or more.

As previously described, it is a help to learn as much as you reasonably can about a present system since the problems and errors like those you have now will continue to plague you as long as you are in business. But you must keep the communication and learning lines open to find out what the future looks like. For instance, suppose that most of the personnel in top management is relatively older and more accustomed to running a business based upon an intuition basis. If the younger people on the way up have recent MBA degrees and are highly oriented in quantitative methods, then it would appear that you would want to employ business techniques that these people would be more comfortable with and would be expecting to see at the time the new system is operating.

To be able to properly plan, the Systems Department must constantly be finding, in a general sense, what the plans of the business are:

1. Are they expanding into new products or geographical areas?
2. Are they cutting back operations anywhere?
3. Are they planning a merger? One large airline went through a complete systems change only to be merged with another line that had already done the work.
4. Have broad objectives changed; is the expected rate of return now higher than had been quoted in the past?
5. Is it obvious that a person with apparent strengths or weaknesses is moving into a certain position?

Of course, management won't always take systems personnel into its confidence and tell them all the things that are about to happen. But by participating to some extent in the projects being worked on, management can prevent time spent in areas which will be eliminated or merged with a good system which is already working. Management should know enough about systems work to steer efforts in the right direction.

Final Selection

In determining exactly which of several alternative major systems is going to be used, it is desirable to get answers to the following questions.

1. What is going to be the approximate cost to operate the system once it has been set up? The answer to this question will be obtained by making a cost estimate used upon the general systems design that had been arrived at according to the guidelines

in Chapter 7. Although none of the alternatives may have a lower cost than the present system, you certainly want to compare costs of the various proposals to their respective benefits.

2. What are the remaining costs in completing the detail design and finally installing the system? You must be careful to make sure that this estimate is not too optimistic. Many people have found that such estimates are only 50 percent or less of what the actual time and cost eventually turn out to be. By the same token, once you have arrived at this figure for each of the proposals, you must realistically determine if the business is in a position to finance that expenditure recognizing that it will be some time before any return will be realized.

3. What percent of the day or of the available hours in the shift or shifts will be required to operate the system? For instance, if management has already established the policy that all work is to be completed in one shift without overtime, you must relate the timing estimates to the number of hours available in that shift. If the estimates appear to take up more than 80 to 85 percent, then it is unlikely that the job can be done on a realistic basis within the capacity of one shift. Many companies have found that if the schedule takes up more than 80 to 85 percent of the total hours available on a 24-hour-per-day, 7-day-per-week computer operation, the computer center quickly gets in trouble trying to do all the work. This is because of all the things that must be done which may not have been realistically planned for when setting up the schedule. Examples of things that will cause the schedule to go amiss are programmer test time, down time, preventive maintenance, and rerun time. This last item sometimes consumes as much as 10 percent of the total time available.

4. How closely have each of the broad proposals come to meeting the specifications? The specifications should include all the objectives as they relate to what should be done and the profit to be gained by doing all those things (benefits minus costs of obtaining them).

5. How close are we now to having the people that will be required to operate under such a proposal? It is likely that some organizations do not have enough qualified people to operate under conditions that a complex system requires, particularly those of a sophisticated computer center. If there are not enough internal people of the caliber required, the company should carefully consider if it can afford to go outside and bring in people who know

little about the company's operation. There may be certain categories of people that can be readily obtained but if all the expertise you need is on the outside, you will need to proceed with caution. In any case, it will take a while to obtain them and train them in the methods peculiar to your operation.

6. How many major changes are required or how much disruption is really going to occur as a result of making these changes? How willing are people to accept change? If the organization is of the type where most of the top people are relatively older but perhaps not at the point where they are about to retire, it just may not be worth trying to make substantial changes now. It may be better to wait until such time as these people have moved out of the picture or can be conveniently put in a place where they are not so likely to hold back desirable change.

In all of the solutions you arrive at you must make sure there is a proper balance of all the specifications. For instance, it would be possible to have a system in which inventory turnover would be very high. This could be done by setting up a distribution system in which deliveries would be made to customers as soon as goods were received. Of course, this would require that you only ordered for customer use when you had a specific customer order and then immediately delivered it in order to have no inventory on hand for yourself. This would give a very low average inventory figure and thus result in a higher turnover figure. But the total gross profit obtained from a few transactions you may have by operating in this manner would not be very high because you would not have many customers and sales.

Another example of how you must carefully balance your specifications is that a very high gross profit rate might be realized by having one or two fantastic sales in a year to a customer who perhaps did not know current prices or who was in a real bind to get the product quickly. As a result of charging an abnormally high markup it might be possible to obtain a very high gross profit rate but again you probably wouldn't have many sales. Your total gross profit dollars would not be high enough to cover all expenses and still provide a net profit.

There is a point where benefits will not continue to rise any more in relation to the increase in cost to obtain those benefits. Notice how this particular concept applies when you shop for a car. As you proceed up the price range by thousands and eventually get into the $25,000 category, the size, ride, speed, acceleration, and other characteristics of a car tend to have a very diminishing return as you would spend additional thousands of dollars. The same thing will happen in a system whereby spending vastly increasing amounts of money may not keep the benefits rising proportionately. For instance, using a college graduate for a keypunching job may not provide any greater speed or accuracy than you may be able to obtain from somebody who has a high school diploma. So in systems design, you are not always looking for the greatest capacity to do the job but are looking for methods that tend to provide the best overall gain after considering the cost of obtaining those benefits.

Questions

1. How can a systems department make sure it doesn't spend a lot of time on an application that may be eliminated in a few months?

2. A company's scheduled day shift runs from 8 to 12 and 1 to 5. How many productive hours can they expect from a clerical worker in that time? Why has a company typically had more productive time from a nonclerical worker?

3. Assume your payroll system is set up to pay people by check. List several different ways of distributing checks to employees, and give any advantages and disadvantages of each.

4. What is meant by the "law of diminishing return"? Give an example how the law may apply to some phase of systems work.

5. Discuss the pros and cons of the following ways of getting letters typed:
   a. Writing them out in longhand for later typing by a secretary.
   b. Dictating to a secretary who types the letter at the same time.
   c. Dictating to a secretary who records in some fashion of shorthand and types them later.
   d. Dictating material into a recording device which is later transcribed by a typist.

6. A college teacher living in Boston who wanted to go to Europe found he could fly directly from Boston at 10:00 P.M. for $275 or he could go to New York and depart at 10:00 P.M. for $225. (Both are round-trip fares.) What considerations should he make as to which would be the better financial arrangement?

7. Suggest a way that a utility such as a telephone or water company might obtain funds other than by borrowing or selling stock.

8. A certain railroad tunnel has a clearance of 16 feet. Some new equipment will eventually pass through requiring a clearance of 18 feet. What is the best way to obtain the additional 2 feet?

9. In studying the computerization of many of its business functions, a large company found that it had enough volume to keep five computers busy. Give two important reasons why the company should get all five computers of the same type. Give two important reasons why all five computers should not be alike.

10. Jones Company is about to get a computer and is trying to determine whether they should keypunch and key-verify cards and then transcribe to magnetic tape, or to key to tape directly. From the following, determine the cheaper way.
   a. Transactions numbering 7500 of 40 characters each are made per day.
   b. Keypunch machines rent for $80 each; verifiers $88 each.
   c. Cards cost $1 per thousand.
   d. It will require 15 minutes of computer time each day to transcribe from cards to tape. The computer is costed at $50 per hour.
   e. The key-to-tape machine rents for $175 per month each. Since the operator will be keying and verifying at the same time, her effective speed would be 5000 key strokes per hour. You may disregard the cost of tape and any allocation of space. Make and state necessary assumptions. Show your calculations as to which is cheaper.

CHAPTER 10

Selling the Proposed System

To a large degree the systems analyst is a coordinator, not a doer. That is, his job exists only to help others do their work in a better way. Thus, he is in the position of being an internal consultant, and his recommendations are presumably subject to acceptance not only by the operating people who are going to be using the systems he designs but also by management. For the good of the company, it is best if he is not so powerful that he can introduce major things without the approval of those who are deeply involved or ultimately affected by them. Hopefully, at this point in the design cycle, management is still vitally interested enough in what is going on to actively participate and see the project through to its completion.

You as a systems analyst may have done an excellent job up to this point, but there is nothing to guarantee that your recommendations will be adopted. You most likely will have to enter into a very effective campaign to sell your proposal because of any of the following reasons.

1. Perhaps the people in the operating departments have considered you to be an intruder as you have gone through the steps of studying their work and requirements. During that process they may have felt that you would eventually go away and leave them alone. But you are now at a point where you are formally recommending that something different be done, and you may have to muster all the power you have to convince them you have something solid.

2. Previously, you or other systems people may have made serious mistakes in judgment. If so, your job is now harder because you must do what is necessary to eliminate that negative feeling and get things back on a positive basis.

3. If some of the people involved have feelings of inadequacy, they may fear that you are about to expose them for what they are. If they have built up large staffs, it may appear that what you are doing is going to break up their empires. Thus, they may tend to dispute what you are doing. You will have to handle this type of situation in a delicate manner to convince them that what you have will work for them.

4. A supervisor or key employee in an area may be approaching retirement, and he may not be interested in learning something new or going through the steps to install something which he will only work with for a year or so. Resistance to change is not restricted to older people, though. Many people just starting their

155

careers are set in their ways. Opposition often occurs because of a lack of understanding.

5. Up to this point all design costs have been charged to the Systems Department. Users can now see that certain operating costs may be increasing and that such costs will be charged to them. They often feel it is a personal reflection upon them if their budgets must be increased.

6. People in departments may feel they are doing all right now and all they need is for the Personnel Department to start sending them more or better people or that they would be all right if the Central Office would just let them alone.

7. Top management will want to see in a general way what you have to present, what it will do, and what it will mean. Also, they may not act positively on your recommendations unless the operating departments have endorsed you first.

You should know ahead of time there may be some resistance, and you will then be in a position to plan for it.

Recognizing that in many cases there may have been some selling involved to get management to hire you as a systems analyst and to permit you to take the time to make the study in the first place, selling a proposal may only involve the gentle art of persuasion to show the following.

1. Something of this type has to be done because of growing volume, continuing scarcity of qualified people, or increasing costs of people as opposed to the costs of automated methods.

2. Your solution is better for your company than the one they may have seen somewhere else (at a competitor's, maybe).

Without a doubt the best approach to follow is to make the new system so attractive that all people affected will want to adopt it without much selling effort on your part. This can often be accomplished as a result of the following.

1. Always make sure you know what you are talking about. Carefully plan your actions so you are right most of the time.

2. Work very closely with all concerned, rather than working in a vacuum.

3. Go into an area with the attitude that you are not an expert about their business but want to learn everything you can from them.

4. Listen attentively to worker suggestions and requests for service, and follow up with positive programs for help.
5. Prove that you have no preconceived ideas about how something is to be done. Everyone will quickly lose faith in you if you go in with a firm solution already in mind.
6. Relate what you are doing to the overall goals of the department and the company (hopefully the people know what those are and now have clear knowledge of where they fit into the total picture). This approach will be very impressive to top management.

Presenting and Selling Ideas

For reasons like those given below, it will be necessary for you as the systems analyst to make various presentations regarding the system you have designed. Obviously the system is not going to be installed first in order to find how generally acceptable it is. Presentations are needed for the following reasons.

1. You will ordinarily be in a position to recommend only, and this is your chance to make your recommendation. Hopefully you will have done such an outstanding job that your ideas are adopted without undue pressure from any source.
2. Perhaps your work has always been good. Then it may not now be so much of a problem convincing operating departments and management in what direction they should move. Perhaps you need to inform them generally as to how the system is going to work, the types of changes that will have to be made, what it is going to cost, what the benefits will be, and a timetable to put it all into effect.
3. The worst situation occurs when people in the organization are not getting along with each other. If the operating people don't think the new design will work, the analyst may have to prove himself. It makes matters worse if top management is pushing everyone too hard.

The types of presentations you may have to make will fall into the following categories.

1. *Informational.* This type would be used mainly just to inform the people. You may be telling or showing what they can expect from the system. This type may be needed to rid fears that people may have about their own jobs or what the future holds.

It would be best if top management would give this type, but they may delegate it down the line.

2. *Instructional.* Telling or teaching those who are going to be operating the system what is necessary to make it work. If this should turn out to be a lengthy assignment, it would be better to develop operating personnel to teach within their own departments. Employees would relate better to this form of instruction, and you would be freed for more creative activities.

3. *Request for information.* Perhaps you need to know something about management's philosophy, goals, or time schedule. This type of meeting is relatively brief, and most of the information is actually flowing to you rather than being presented by you. Or you may tell what you have learned so far and ask for a verification of it.

4. *Request for a decision.* This may involve going to management and telling them about a certain aspect of the operation. Then you request perhaps a certain budget amount, ask for approval to do a certain thing, or may even ask just for the authority to continue. Or it may be to ask for approval of your whole package and authorization to put it into effect. It would be nice to expect an immediate decision, but quite often management will need time to assess your request.

You must recognize that anything you do in your presentations will have a direct bearing on the opinions people have of your ability, regardless of how good you may actually be. For instance, if your writing and speaking include such words as "ain't" and "you was," the audience will have a natural inclination to suspect the quality of your systems work may not be up to par, either.

In getting ready for any presentation you might use as a general guideline your own knowledge of the type of thing you would expect or require if you were to be on the receiving end. From my own experience, I think the two most serious faults of presentations in general are the following.

1. They do not have the correct detail for the respective audience.
2. Too often the material does not adequately indicate any of the disadvantages or the pitfalls. It is reasonable to expect some disadvantages in any approach that may be taken, and recipients are more likely to believe and put faith in you if you have revealed both sides of every story.

You must carefully relate everything you do to the
people it is presented to. Generally, management is no
in many of the details. They want to know if it will wo
will rely upon operating people for feedback on this
soon you can get it going, what it is going to cost, what
will be, and perhaps how similar it might be to anything that competitors have. So it is best not to give them a 250-page report or a
three-hour speech as they will immediately question your compe-
tency.

Even when formally presenting ideas to operating people it
is not necessary to go into details on everything. Much of this can
often be done in private with the interested people. At the formal
presentation you might want to cover the details of only one segment of the overall system. But you would want to have details
available on all the others in order to show and back up what
you've done, and you would want to be able to answer questions.

You should try to use all the appropriate audio-visual devices
to show your material in the best possible way, since much of that
will be preferable to a plain speech. These devices would include
charts, flow charts, graphs, slides, and even movies in certain cases.
Of course, it is going to take time and money to prepare properly,
but it is something you need to do only once for a system that might
be used for years. In this regard, the two major problems to overcome are those related to sight and sound.

1. I have attended many presentations where the leader had
   charts on a stand or had projected a slide on a screen and
   has then prefaced all other remarks with "You probably
   can't see this in the back because it is so small."
2. In a room which is not adequately designed for proper
   sound distribution, a voice projection system must be
   employed. So many of these sound systems work poorly
   and snap and crackle to the point where the speaker often
   gives up using them and just speaks as loudly as he can
   with the hope that everyone can hear.

Both of these problems can be overcome by proper planning
and a dry run in the place where the meeting is to be held. If
twenty-five people are to be brought together for just one hour,
that is twenty-five man hours of time taken away from other duties,
and it is important that you do enough homework to be able to
present everything in a manner which is as pleasant and understandable as possible.

Use of the Negative Sell

Occasionally you will find yourself in the position of having to play down or justify a negative position on a proposal that some management person may have made. Perhaps he has seen a very sophisticated system in operation at some other point and believes this is exactly the thing that your company should have. But you must be realistic and recognize that just because another operation has a particular thing does not necessarily mean that your employer is ready and able to use such an item.

There is a large industrial company that maintains a fleet of airplanes which is larger than the fleets of some of the regional airlines in this country. A few years ago one of the executives, who used the company flights frequently, went to the president and requested that a computer on line real time system similar to those of the major scheduled airlines be adopted. He pointed out that he could always call his favorite scheduled airline and they could tell him precisely when a plane was leaving that would be able to take him to the destination of his choice. When he called the scheduling people of his internal airline division, he found it very difficult to get a firm commitment as to when a plane would be going to a particular location. The president asked the Systems portion of the Data Processing Department to review the operations of the Airline Division and determine whether an airline type of scheduling and reservation might be put into effect within the company. During the course of its facts-gathering phase, the Systems Department found a number of things about internal airplane scheduling and operation that made it difficult to develop something comparable to what the scheduled airlines had in operation. Some of the things which made their operation substantially different were these.

1. The company had precise policy restrictions as to how many and exactly what officials could fly on the same plane at the same time. For instance, the chairman of the board and the president were not allowed to be on the same flight. Only a certain number of management people were permitted to fly together. Because of such rules, the company found that it occasionally had to use three or four planes to take people to a particular point when in fact any one airplane was big enough for all of them. But this meant that planes may not be available to fly to some other destination. It was determined that this policy would not be changed.

2. One of the major reasons for having the airplane fleet was to have the planes on call during the mid-evening hours to haul components among the various manufacturing plants of the corporation so that no plant would have to be idle the next day because of the lack of materials.

3. Some high-ranking officials refused to fly on certain planes. This obviously made scheduling much more difficult.

As a result of their brief study the Systems Department determined that this particular operation could not be adequately helped as a result of using a computer. There just weren't enough consistent transactions to make a sophisticated computer system worthwhile. Incidentally, this is the only case I have ever heard of where a computer feasibility study resulted in a negative recommendation.

Inventory Control

The easiest job you will ever have of selling a proposed system is when your general design has been done in such a way that the expected benefits are obvious. In some of the examples given in Chapter 7, there are a number of suggested changes which, if adopted, would seem to have an immediate and substantial effect upon profits. Some would perhaps improve customer service, which would eventually have an effect upon profits. If you are so fortunate in finding such obvious problems in other systems, you may find that management will wonder how they got along on the present systems. The best selling is of the type where there is no selling at all but where management is so happy with the design that they readily buy what you have proposed.

In your presentations it would be desirable to show charts and graphs and perhaps simple flow charts of some of the main concepts involved in your general systems design. Since you will perhaps be making several levels of presentations, perhaps one to some of the top management of the company and at least one more to the various managers or supervisors of all affected departments, then you will need to have several versions of presentation at hand. You may have only from fifteen to thirty minutes with top management, but conceivably your presentation and justification to the working managers may take half a day or many sessions.

It is entirely possible that portions of your inventory control system may be put into effect without involving top management.

For instance, it would be quite possible for the Inventory Control Department to change their method of calculating reorder points and reorder quantities. In many cases such action would be internal to them only and would not require any approval by outsiders. But if there is any possibility that the new procedure might occasionally cause an inventory item to be out of stock and thus perhaps result in a missed sale, and if the manager of the department is very conservative and has had a long history of never being out of stock, it may be difficult to get him to budge at all on this point unless certain pressure is brought to bear on him by others. Obviously, the person in that position is not going to make any kind of move if he sees the possibility that some destructive criticism may eventually come his way because of making such a decision. The same general point applies to the thought of eliminating the double processing of orders against the perpetual file.

For the most part, though, systems work will involve applications which deal directly with a number of different departments in a company. This is commonly referred to as activities "which tend to cross departmental lines." Regarding the inventory control system that has been discussed in this book, consider the following.

> For various reasons the Order Entry Section of the company has always handled back orders. Apparently, it has been done this way because back orders are problems that are most closely related to the Sales Department as they have been the ones who deal directly with customers on such matters. Since Order Entry personnel report to the manager of the Sales Department in this company, back orders have been handled as they are.

Because of the way in which back orders are handled, that is, because Inventory Control notifies Order Entry when there is sufficient quantity on hand to ship represented by a back order, it takes at least two days from the time the material is known to be available until the paperwork is sent to the Shipping Department to tell them to ship items on a back order. If the new systems design would call for Inventory Control to hold those shipping documents and then forward them to the Shipping Department when the material is available, the following would happen.

1. The material can be sent on its way to the customer an average of two days sooner.

2. Billing the customer and ultimate receipt of his payment will occur sooner.

3. Quicker shipment may tend to eliminate some of the cancelling of back orders that is always common when the customer can't get his order completed within a reasonable time.

It is likely that the sales manager may think very little of the proposal that control of back orders be taken away from him and be placed in the Inventory Control Department. The best you can do on this point is to show with facts what effect such a change is going to have on the overall operation. Once top management is convinced that a particular change is going to be worthwhile, it is quite likely that they will want to convince him one way or another that the change is in the best interests of the company. If it turns out that the sales manager reacts unfavorably to this, his only defense is going to be some kind of reasonable proof that the proposed change will not work. You might suggest putting the new procedure into effect on a trial basis with respect to a couple sample items and try it with employees who are known to work reasonably well under such a situation. As a last resort, you might want to suggest that he move one of his own people to the Inventory Control Department and operate it in that fashion.

Ultimately, you want to make sure you have furnished your best efforts to reduce the prevalence of back orders rather than just devoting all your time to figuring out how to handle them.

In reviewing the work that has been done so far on this inventory control system, it is somewhat difficult to point out disadvantages or pitfalls with the possible exception of these.

1. It is going to take some time and money to put the changes into effect. The cost seems to be minimal in this case, but it will have to be weighed against expected benefits.

2. There is a possibility that some sales could be lost as a result of allowing the levels of certain inventory items to be lower in the future than what they have been in the past. However, if you have done your statistical and forecasting work properly, you will be able to show the net overall effect on profits as a result of occasionally running out of an item.

3. Another significant item that must not be overlooked in any major change relates to the possible bad feelings that people may eventually develop because their system has been "criti-

cized." However, you must recognize that the system has gotten into the condition it has because some person or persons had not previously done the best job that could have been done at the time. This is the very reason why many corporations have suggestion programs which pay out millions of dollars each year to employees who come up with money saving ideas. In many cases people are not eligible for suggestion awards relating directly to their own departments on the basis that it is their job to solve their own problems. But if someone outside the department can come up with the idea, then he should be rewarded for doing so. Realistically speaking, it is because someone is not doing his job completely that has caused companies to make rewards for ideas from others. You will just have to be positive on this point and try to prevent any ill will.

Questions

1. Why might top management lose interest in a project before it is completed? What can be done to retain their interest?

2. Why might you have to fight all the way to get your employer to agree to a new system?

3. Briefly outline the steps that you should follow to make sure a slide presentation with questions and answers to follow goes well. Name several things that could happen that might reduce the effectiveness of such a program.

4. Why would you possibly have to prove that a system working well at your main competitor may not be any good for your company?

5. Why should you always present some of the disadvantages of a proposed system?

6. Suppose your company is not borrowing any money and therefore has no interest expense. Prepare a brief statement proving that the company can save money by reducing the amounts tied up in inventory and accounts receivable.

7. Suppose a telephone company advertises that a push button phone is three times as fast as a dial phone. What is the approximate time difference in making 20 calls?

Other Readings

Hodnett, Edward, *Effective Presentations*. West Nyack, N.Y.: Parker Publishing Company, Inc., 1967. Gives principles that will help to improve both oral and written presentations.

Neuschel, Richard F., *Management by System*. New York: McGraw-Hill Book Company, Inc., 1960. Chapter 15 covers the presenting and selling of ideas.

CHAPTER 11

Installation

To arrive at this point in practice you would have successfully sold at least part of your plan and received the green light to move ahead. Up to now you have been dealing with concepts, generalities, and ideas. To a large extent the hardest part of your work is just about to start. You have done about everything so far yourself, and it is now time that many others start participating and cooperating to a high degree. You will be the one to cause this to happen, along with a necessary boost by management. You will not have completely proven yourself until installation is made and benefits start accruing according to the plan.

Despite how enthusiastic someone may be over a system, he will tend to lose interest unless he can see that solid progress is being made. As soon as the approval has been received to proceed, you should establish the direction that implementation is going to take and make sure that progress is shown quickly. While you were selling the system you should have presented a general idea of what would be involved in installing and how long it would take. In fact, your design work up to that point wasn't complete unless you had provided for it.

Depending upon the complexity of the system there will seem to be a thousand and one things to do. It will take a great deal of coordination to get them done properly, and if you are not careful, many of the future benefits can be dissipated at this point by not doing them correctly. In addition, the upset to routine operations can be fantastic.

All through the steps of analysis and design you should have been making notes of the things which would have to be done to effect complete installation. In general, a major system would involve such things as:

1. selection of people
2. training people
3. providing for displaced people
4. order equipment and services
5. parallel operation
6. followup

Obviously these don't have to all start at the same time, they don't all have to be completed at the same time, and there are going to be many people who must be actively working together to get the job done. And management may need to be available to make decisions and help clear up problems all along the way.

Selection of People

If the nature of the system change is such that people need to be chosen to perform new tasks, the desirable approach for you as the systems analyst is to get into the position to recommend those people that you feel will do the best job. As a result of spending a great deal of time among these employees during the design stage you are in a very good position to be able to make such a recommendation. However, it is possible that because of the relative power of certain management people, others may attempt to make all the rules regarding the people who are going to be chosen. There are really four different groups deeply involved in employee selection: top management with ultimate responsibility to stockholders, operating departments to make it all work, the employees to be selected because they must want to do the job and be compatible with their co-workers, and the systems analyst to coordinate it all.

One of the better approaches to take when you have been provided with names of present employees is to deal very carefully with any situations where a person's supervisor is intensely interested in transferring his employee to the new effort. It sounds too much as if the supervisor is trying to get rid of the person, and if you accept the person without investigation, you may inherit an employee who is not so desirable to have. Many personnel studies over the years have tended to prove that most work problems are due to the inability of people to get along well with each other. But in any particular situation you might also find that a person is just not competent for the job. So you should try to obtain those people whom the supervisor says he cannot possibly part with because these are the ones who are better suited for the new effort.

There will be times when someone may have had a preconceived feeling that a particular individual was an immediate heir to one of these new positions. If you have any doubts as to that person's qualifications, you may need to resist such efforts with all reasonable means. There is certainly no specific advice that can be given on this point except that you must try to do what is best for the company. In the final analysis, everyone must recognize that the Systems Department is acting only in an advisory capacity, and it is up to management to make decisions regarding people who are chosen.

In many cases, the new system is not going to require something drastically new or different from the present one. In the sample recommendations for our inventory control system, there is not that much that has been added in the Inventory Control Department or any of the other areas. Either the activities are still being done but at a different time or different place, or an activity is being eliminated. So it may be that you would not have a personnel selection problem at all.

Training People

No matter what changes are needed to put a new system in operation, there is going to be a minimum of training involved. This training may vary in length from a few minutes of instruction in the office to a program where a person would have to attend a special course for weeks in order to learn and apply a new technique. The computer programming, repair, and operation fields have had plenty of activity of this type where a complete exposure may last as much as several months. For a nontechnical job that would be required in the operation of a typical system though, such a long course probably would not be necessary. Much training of this type can be done on the job as opposed to sending the employee to a formal classroom atmosphere.

A very worthwhile way of doing much of the training is by the use of the programmed instruction (PI) method. Figures 11-1 and 11-2 show the basic approach that is used. The nature of this

Figure 11-1.   Sample of programmed instruction.

Inventory turnover is calculated by dividing the cost value of material sold by the average inventory that was on hand during the year. If $20,000 worth of material were sold during the year and average inventory was $4,000, turnover would be

_____

* *

Ans. $20,000 ÷ $4,000 = 5

(Programmed instruction often takes the above style where a paragraph or so explains a point. Then a pertinent question is asked and the reader is to place his answer at a specific place. The correct answer appears beneath a row of asterisks. The student thus finds out right away if he is absorbing the material. Specific instructions at the beginning of the course would tell the student to cover the page below the asterisks with a heavy sheet of paper so he could not see the answer before he selected his own.)

Figure 11-2. Sample of programmed instruction material—manual recording for an optical scanner.

FOLLOW THESE RULES FOR PRINTING THE NUMBER "TWO":

A. Make a simple hook — no fancy loops or curls.
B. Make the bottom flat.
C. Remember the size rule — the "TWO" should almost fill the white box without getting into the border.

PRACTICE DRILL

Look at the practice sheet. Trace the "TWOS" on Line D with your pencil, filling the line with "TWOS" like the model.
Then follow the "ZERO," "TWO" pattern on Line E.

LINE

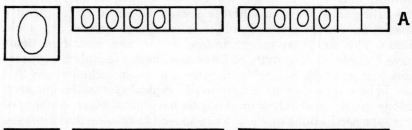

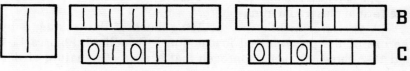

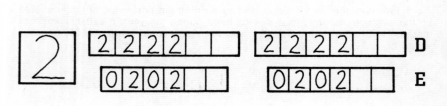

method of training is to have the student proceed at his own pace without having a teacher lecture for the purpose of telling him everything. However, it is desirable to have some one person available in an instructor capacity to help with questions for which answers may not directly appear in the text material. The method should certainly save a lot of time and money that would otherwise be spent on the normal teaching approach, providing there are enough students over which to spread the high cost of preparing the material.

In an educational effort it is necessary to avoid such general terms as handles, processes, and studies. Statements regarding what people are to do must be specific. Instead of instructing a person to "handle" a form, you must indicate that he should perhaps pick up the paper, look at a specific place for an amount, multiply that amount by a value, write the answer at a specific point on the paper, and then stack the papers on the right side of the desk and at four times during the day carry them to the supervisor. Notice that a word such as "handles" is not a very effective term to use in telling a person what his job is going to be.

Regardless of the method of training that is used, it is most desirable that the company arrange adequate time on the job while the person is getting paid at his normal rate. The company will probably profit in the long run from this approach rather than trying to force the person to get the training on his own time.

One of the most important aspects of training is to make sure that the person in charge of the program does not falsely assume that the employee understands what he is doing without making a reasonable check to see that he is progressing all right. Several years ago I assigned a data processing class a term project that involved designing a system to maintain a name and address file for 50,000 customers. One of the requirements of the project was to provide in the systems flow for adding about 5000 new customer names each month. Shortly after receiving the assignment, one student came to me and said he would not possibly be able to complete the project because he did not have enough time to punch the 5000 cards each month for the remainder of the semester. Obviously nothing in the assignment pointed to the fact that he had to physically prepare 5000 cards; he only had to provide in his system for that many new items and calculate the estimated time and cost involved in processing them.

This example merely illustrates the point that it is so necessary to keep checking on the progress of people to make sure they understand what they are to do and to be able to move ahead accordingly.

A big part of the training effort that will go a long way to help get the system underway soon and on a reasonable basis is to make exactly sure that people know why it is that their position is required and how their work dovetails with other departments of the company. The relatively small amount of time spent on this effort will more than pay for itself in terms of increased employee understanding of the total picture.

Providing for Displaced Employees

In those cases where operating steps are only changed somewhat, there may not be any people who have lost their positions as a result of the systems change. Obviously in that case, you are not going to have any displaced persons unless you have brought in someone else to take over an existing person's job.

But quite often, due to a major realignment of jobs, reducing the complexity of certain steps, cutting out certain steps completely, or mechanizing a particular activity, there may be people who will not have the position they formerly held. This is normally not due so much to the fact that there are fewer people needed but that jobs have been created that require people with a generally higher level or different type of skills. The best thing that could happen here would be that the people holding all the eliminated jobs would all be qualified enough to assume the different positions in the new system. However, because of the qualifications of the people and perhaps because of their general attitude or ability to get along with others or for one of many other reasons, it may be most undesirable to move some of the existing people into the new positions.

Occasionally there will be some employees who, from the company standpoint, should be given early retirement or perhaps immediate separation from the organization. Some companies do take the hard-nosed approach, fire them, and then blame the whole thing on the computer, the consultant, or the Systems Department. You are certainly in an awkward position at this point, but you must remember it is a management decision in the final analysis as to who is chosen and who may have to be let go. In many cases, the company has had many more employees than they really needed, and if they are to have true savings from a systems change, the only thing to do may be to get rid of people in order to eliminate high costs. Such things as retirement and resignations can be counted upon to provide for some of the attrition that is needed in order to reduce the number of people, but those particular occurrences cannot be relied upon to do the job in every case. Worse yet, they do not always assure that the most undesirable people are the ones to leave.

There are occasions when such things as company policy, internal politics, or union contract rules provide that an employee cannot be relieved of his job or moved to a lower ranking job. Once you understand the rules with which you must live, it is probably best if you will try to avoid any personality conflicts and perhaps try to stick to the system in terms of working with the total number of people needed instead of specifying people who should be on the way out.

Order Equipment and Services

Enough work should have been done prior to selling the new system so that equipment selection, some basic forms design, and supply specifications had already taken place. In the event that this has not been accomplished, those details must be taken care of now so that orders can be placed with the proper vendors to assure delivery of all those items according to your schedule. (Note you want delivery according to your schedule, not to accommodate the vendor.)

Once the proper items have been selected, the two important points to be careful about are that purchase order terms are completely known and agreeable to you and that you properly follow up to make sure of the delivery schedule. Of special interest in purchase order terms are such things as follows.

1. Provide for a cancellation or penalty clause if delivery is not made within a specified period of time. If the vendor is late on delivery, the time when you will begin receiving benefits will be delayed.
2. Written statements as to any help or services that would be provided as part of the purchase price or the specific cost if services are to be given as a separate item.
3. Specific knowledge as to other requirements such as special electrical power or air conditioning must be known.

You will also need to carefully follow up your own procurement operations to make sure they do what they are supposed to do. A company ordered a computer and was told by the manufacturer it would be the user's obligation to obtain a special electrical plug that would be fixed to the machine for insertion into the wall socket. The specifications for the device were turned over to the Purchasing Department but they failed to place an order for it. When the computer arrived it was learned that the item had never been ordered, and the computer could not even be plugged in. You certainly don't want to get the reputation of one who breathes down everyone's neck, but you must do what is reasonably necessary to make sure everyone is doing what must be done.

Parallel Operations

The purpose of a parallel operation is to run the new system and the old system at the same time in order to determine that the new system has been designed and installed all right and that it will produce the required results. This does not necessarily mean that the new system is completely installed at one time; perhaps only one segment of the total or only that portion for one of many branch locations is put into use on a test basis. When that segment has been proved to be all right, installation of additional segments or locations can take place on a planned basis until installation has been completed. When it is reasonably proved that the new system is working all right, then the old one can be completely abandoned.

This is not to suggest that a parallel operation will always be used. Here are some cases where it wouldn't be used.

1. Where this is a new system doing something that had never been done before, there is no parallel operation. For instance, using computers on space flights is the only way the calculations have ever been made; the computer did not replace some other way of getting the job done. Simulated runs in laboratories were used long enough to make sure the computer could do the job properly.

2. Management may feel the system has been so completely debugged and proven elsewhere that it is not necessary to operate in parallel. For instance, this could be the case if the business has chosen to adopt a carbon copy of something that is being done in another organization. However, it must be recognized that it is going to be your inexperienced people trying to make the system work for you, and you are taking a risk that it may not work effectively right from the start. You may want to recommend that no parallel system be operated, but management should be aware of any pros or cons, and they should be given the opportunity to share in making the decision.

3. Quite often you will not have the facilities to operate two systems at once. This would be especially true of a physical system where you may have room for only one assembly line or where you may have had to trade in the old equipment to get the new equipment. But if a company is replacing one computer with another, it is unrealistic to plan on wheeling the old one out the same day the new one is brought in.

4. You may not have enough people to work both systems. One company solved that problem by hiring people expressly for the purpose of operating the old system while existing employees were switched to operation of the new one. Then supervisors had the unpleasant task of releasing the newest employees; the people who hired them had not been candid enough to indicate the work would definitely be of short duration.

5. The company may have so carefully planned and trained for the new system that they can install it on a wholesale basis and remove the old one immediately. It was reported in 1965 that seven savings banks in Eastern Massachusetts converted to an on line real time savings account system without any parallel operations at all. According to published reports all seven institutions were able to balance every account to the penny on the first day of operation. One reason why they chose not to go to a parallel operation was the difficulty of obtaining enough people to operate the old manual method and the new computer method at the same time. Apparently the respective managements of all the institutions had seen enough of that type of system in operation in other banks that they were willing to trust the reputation of the computer manufacturer and the systems that had already been installed. Substantial training had taken place with all tellers so that they knew precisely what was required of them in order to operate the system.

6. The present system may be so undesirable they just want to get rid of it.

There is a great difference of opinion as to how long a parallel operation should continue. In general it should be continued until such time as you are satisfied that the new one is doing what it should. In a negative sense you would operate until you were convinced that the new one would not work and you would decide to get rid of it. There must be a practical point where you are willing to admit you may have made a mistake and revert to the old system. A department store in Detroit got rid of its computer and went back to manual billing because the computer "didn't have a heart." As with other computer systems that have failed, this one had not been properly planned, and due to many billing errors too many customers were closing their accounts and buying from competitors.

Follow-Up

Follow-up is a series of procedures that should be used some time after installation has been complete and the system has had enough time to function under normal conditions. It may be a few months after installation or even a year or so later. The basic purpose is to review operations in light of what had been planned.

Whereas the purpose of parallel operation is to hold on to the old system until the new one is operating smoothly, the follow-up described below is much more complete in that actual costs and benefits will be carefully reviewed. Although in practice the study may often be delegated to members of the Systems Department, it should really be a direct responsibility of management to see that it is carried out according to their directions. It is possible that systems people would have some bias because much of the system was their own creation, and the follow-up should realistically review what everybody has done in getting the system designed, installed, and operating. Allowing members of the Systems Department to make the review now would be a case of them auditing their own work. For this reason it would be desirable if this review were performed by a group such as the Internal Audit Department or at least by someone other than the one who had primary responsibility for it in the beginning.

The follow-up may be as sketchy or as complete as management wishes to make it. A brief review of the income statement and the balance sheet with a decision that they are in good condition might be enough reason to think most systems are functioning reasonably well. Perhaps management is generally satisfied with progress or maybe they don't want to allocate any resources to the review now. If the system were a particularly difficult one to install or if there were people who were especially difficult to work with, there may be no one who wants to tackle it again. In fact, any kind of formal study now may cause people to infer that something is wrong and that someone is going to stir things up again.

A good reason to make periodic studies is the dampening effect it has on everyone; if people know there are never any follow-ups, they may tend to become a little lax in their work. If they know there are going to be reviews later, they will have an incentive to do a better job, much as a student may become motivated by tests and unannounced quizzes to expend some effort on a course.

The follow-up itself will be concerned with some or all of the following.

1. Costs of the new system and what the costs were projected to be are compared. Since the expected costs of the new system were compared to the actual costs of the old system and approved, it would serve no useful purpose to compare new system costs to old system costs. What is being done now is to find out how far the new one is off target, and that means comparing present costs to the budget.

2. A comparison of actual accuracy to the acceptable limits which were designed is made. You will recall in earlier discussion that absolute accuracy is impossible to attain, and costs rise swiftly as you try to reach that point. So a level of perhaps 98 to 99 percent was accepted as the goal. Of course, a higher level will be happily accepted if that should occur. It will occasionally happen that errors won't be as persistent as you originally thought they might be.

3. Checking with operating people is done to see if they are getting what they should. Although at the time the new system was put into effect, those people may have been getting everything they required, for some reason something may have been cut out because it seemed too much trouble to provide. Or it may be for some other reason that the system may have deteriorated. This review really checks on the operating people. After a system is installed workers often develop shortcuts, and such practices could have eliminated the preparation of some information. Maybe the recipient has not seen fit to do what is necessary to correct the situation. Also, because of the dynamic nature of business operations, there may now be some requirements that had not been necessary in the original design, and a reasonable effort should be made to see if they can be satisfied.

4. Reviewing the adequacy of training, employee selection, and the general suitability of the people in the various operating areas is necessary. The reason for doing this is not to gather information for fault finding or destructive criticism but for an explanation of why things are not as they should be and to provide a base to set up a program for improvement. Since the success of any system is directly related to how well people are doing their jobs, this is a key point. As a result of an honest effort of fact finding, some

major industries with previous records of low pay scales have re-formed and now are realistically competing for talent with attractive salaries. You may eventually conclude that your projections were based upon using talents that can't be realistically obtained at the right price. As a business grows it may be that its percentage of well-qualified help declines. If this is happening and the trend can't be stopped, the business may have to redefine its goals bringing them to a lower level when it can't get the people it needs.

5. Take an attitude survey of employees. This may be done in a completely anonymous way by asking employees how they feel about the company in general and specifically about their supervisors, their work, etc. If this is done in the proper way, it can reveal a great deal concerning the actions of the people.

Scheduling and Controlling Activities

You can be sure that not all the people involved know what is expected of them. So there will have to be some guiding force which will give adequate direction throughout the installation phase.

Conceivably the analyst could do this in a haphazard manner, but he will be on much safer ground if he will prepare a professionally styled schedule, get people to reasonably agree to it, and publish it for all those involved to see and understand.

In the event that the systems changes are major and of the type that will require an extended period of time to put into effect, it is desirable that progress reports be issued to concerned persons to let them know the current status of things. Certainly any benefits that accrue as a result of the new system will only come about as the changes are put into effect, and the longer it takes to get the new system in operation the longer will be the time lapse before the benefits will be realized. Progress reports can probably best be made from the standpoint of preparing an analysis to show how actual progress is moving along in relation to estimated progress. In those cases where progress is not reasonably up to date or in those rare instances where it is moving along better than planned, it is necessary to generally indicate the reason for any variance. Of course, any progress which is not up to par is going to probably involve personalities, and perhaps some explanations should be made as to what a person is doing or not doing that is affecting the program. Notice how desirable it was back in the planning stage to get people to agree to the schedule of conversion because now it becomes much easier to put the pressure where it is required in order to get the program back on target.

Some people will become very resentful of having reports prepared that suggest an individual may be doing something that is impeding progress. But we must all realize that everyone's work is subject to some form of criticism or review by somebody higher up the line; corporation chairmen and presidents, members of Congress, and even the President of the United States have been subject to such review. Management is going to have to follow this situation carefully and if they are really interested in getting the program moving, they must analyze the progress reports and take whatever action is necessary in order to assure that things move smoothly and on time. Of course, if management has taken a hands-off approach all along the way, then it is likely that installation is not going to take place as had been planned.

At the time of conversion to a new system, you must be ready to handle all types of problems and perhaps put in long hours. This might be necessary to make sure that the system keeps operating in some fashion until the new one is functioning well. Many companies at this stage of development have gone through a period when no customer orders are filled, no bills are sent out for shipments that were made, and perhaps the system was so bad that money that did come in may not have been deposited to the bank account. This is a time when a great deal of understanding is required. It is similar to what the management (and fans) of an expansion baseball team must do after it has obtained all its players from the expendable portion of the rosters of the other teams presently in the league.

It would be nice if the conversion were so well planned and everything went so well that no problems developed at this stage, but there are a lot of activities that must be carefully controlled. You are very likely working with many outside vendors over whom you have no direct control. You are also now finding some of your own deficiencies and learning for the first time of the things which you thought were understood and agreed to but now are not properly jelling. At this point, you and everyone else must be alert enough to determine when you have more than what can be handled. You must be flexible enough to make necessary changes in the schedule if that should be the effort that is required in order to help get things straightened out.

Questions

1. Who has ultimate responsibility for selecting people? How much weight should be given to the systems analyst's recommendations?

2. What is programmed instruction? Give several reasons why it has become so popular. What is the greatest educational expense that it tends to reduce?

3. What will have to be done to make sure an employee understands his assignment?

4. Why are there likely to be some personnel problems even though there may be as many people needed in a new system as there were in the old one?

5. What is a parallel operation? How long should it be in effect?

6. Who should review a system to see how well it is performing? How qualified is the systems analyst to perform that function?

7. Suppose a company knows it is going to eliminate ten people two months hence. The company has a very liberal termination policy in terms of a direct payment. From a systems standpoint, what are the implications of waiting to inform the people at 4:30 on their last day versus a one-month's notice of the effective date?

8. An old system cost $100,000 a year. At design time, the new one was projected to cost $105,000. The new one is actually costing $107,500. What figures should be compared now for the purpose of determining relative costs?

9. At what general time of the year would it be best to computerize inventory control? Payroll? Why?

Other Reading

Neuschel, Richard F., *Management by System*. New York: McGraw-Hill Book Company, Inc., 1960. Chapter 16 gives a very complete discussion of the problems involved in installing the changes that have been approved.

CHAPTER 15

CHAPTER 12

Other Systems Techniques

The topics covered in this chapter represent ideas that you can put to good use as you work on systems problems. The coverage here is just enough to let you know what the technique may do for you. Each is a complete subject in itself, with enough material available to be a complete course.

Solving Data Transmission Problems

A considerable portion of the total data processing and systems effort is going to be involved with sending data from the points where transactions occur to processing centers and then on to the ultimate user of the information. So much data transmission is necessary because you can't have processing facilities at all points where transactions occur or where the information is eventually used; processing points must be centralized because they are so costly. Data transmission takes place in two major forms. One is sending a physical document, and the second is sending electronic codes that represent the data.

Systems analysis requires that you make a careful study to determine whether it is the document itself that must be physically transmitted or whether it is permissible to send a representation of the data from one point to another. The major means of physically transporting documents within the confines of an organization is the ever popular hand carrying method. Less frequently used methods involve conveying equipment where volume between specific points is very heavy; perhaps people on bicycles or roller skates are used where operations are widely spread. The use of the U.S. mail system is by far the most popular when the data is to leave your own organization or when the distribution points are geographically dispersed.

In terms of electronic transmission of data, the two common ones are the use of telephone lines and microwave. Electronic transmission requires that an input device accept or read the data; the data is then converted to an electronic form and sent to the destination point. At the destination, the electronic form is decoded and converted through an output device to the same or perhaps a different form than it existed at the input. The input and output devices are collectively referred to as terminals. Examples of input devices are card readers, typewriters or other keyboard units, badge readers, and tape readers. Examples of output devices are line printers, typewriters, card punches, and audio response (voice).

A computer may also serve as a terminal. Use of a computer in this manner has created the saying "computers talking to each other."

The method which is eventually chosen to move data can only be arrived at after having determined exactly what data must be moved, the form in which it must appear when it gets there, and the form in which it was at the origin. In a system such as a bank checking account operation, where the banking industry is presently committed to returning checks to the drawer of the check, data transmission obviously involves sending the checks through the processing cycle. In those inquiry systems where people are trying to find out the status of credit or inventory accounts, a verbal or audio response is satisfactory and data may not have to move in a physical form in either direction to get the job done. You must carefully determine whether or not a physical document must be moved to a new location, because the cost of the operation, the number of errors made, and the resulting time lag will all be factors in deciding whether physical or electronic transmission means should be used.

A recent form of electronic transmission which is gaining in popularity is referred to as facsimile transmission. To use this method involves a device which can read a printed document at the origin point and convert all the written information to electronic code to be sent over a telephone line. At the destination point, the code is converted back to a printed document. This particular type of transmission has a great deal of use in systems where it is necessary to immediately receive data such as a copy of a contract showing its signatures or perhaps a picture of a person for identification purposes. The major drawback of this form of transmission when it was first developed was its cost and the fact that it took about six minutes to transmit one page of data.

As soon as you begin to explore the field of electronic data transmission, you find there is a wide variety of solutions to every problem. It requires a certain amount of experience to wade through the breadth of vendors, devices, error detection methods, rated speeds, and costs. It is necessary to have a good glossary handy until you have mastered the jargon of the field.

PERT

PERT is an acronym (an acronym is a word that is composed of one or more letters from other words) which is formed from the first letters of the words Program Evaluation and Review Technique. PERT had its formal beginning in the Department of Defense where the technique was credited with helping to speed up the development of the Polaris submarine by approximately two years. It has since enjoyed a great amount of success in any type of program which involves a number of interrelated activities that must be completed in a given period of time. To the systems analyst, PERT has a great deal of application because it can be used to help schedule and control all the activities of getting a system designed and installed.

In presenting an example of how PERT would be used, it is necessary to describe some terms and see some basic guidelines. An activity is a job that is to be performed, such as painting a building or writing a computer program. An activity will take time to perform, and it will cost some money to complete. An activity will be shown as a line with an arrow pointing in a general direction to the right.

An event is a point in time when an activity can start or when an activity can end. Thus, an event itself does not consume time nor does it cost money. An event will be represented by a small circle.

The following illustrates the interrelationships of activities and events.

1.   (A) ———▶ (B) ———▶ (C)

Example: Walls must be put up before they can be painted.

Meaning: Activity AB must be completed before Activity BC can be started.

2.

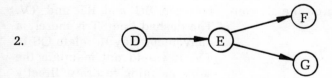

Example: A roof must be completed before the TV antenna is put up or the rainpipe installed.

Meaning: Activity DE must be completed before either EF or EG can be started. EF and EG do not have to be started at the same time.

3.

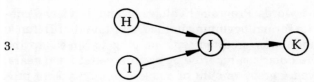

Example: The excavation must be completed and the materials received before construction can begin.

Meaning: Both HJ and IJ must be completed before JK can be started.

4.

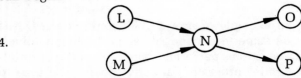

Example: The computer must be installed and air conditioning must be working before you can test programs or run the ones that were debugged previously.

Meaning: Both LN and MN must be completed before either NO or NP can be started. NO and NP do not have to be started at the same time.

5.

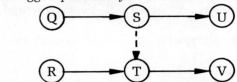

Example: You must start the engine of a car before you turn on the air conditioner or drive away. You must fasten the seat belt before you drive away. But you do not have to fasten the seat belt to turn on the air conditioner.

Meaning: QS must be completed before SU can begin. RT must be completed before TV can begin. Furthermore, QS must be completed before TV can begin. There is no direct relationship between SU and RT and TV. The dashed line ST is merely a convenient way to relate QS to TV. It would not maintain the same meaning to draw directly from Q to T or from T to U.

In terms of getting started, it is necessary to obtain a list of the activities that would have to be performed along with details as to priorities. In a systems analysis environment you should be the one who would know the most about what had to be done; in a project such as construction of a skyscraper, the PERT analyst would need to obtain the data from the construction supervisors. You would then gradually begin the preparation of a PERT diagram similar to that which is shown later in this section. After the activities themselves had been identified, you would do what is necessary to place a time factor on each of the activities. There would be a different approach used to obtain time factors for common activities as opposed to those for perhaps a research job that may never have been done before. In the event that a particular activity is something that has been performed before and thus can be reasonably expected to take the same time again, you would merely determine or obtain the best estimate of the time it would take to complete it.

In the case of the research type effort, the typical approach is to get three time estimates from the most knowledgeable persons and calculate an average. A common method is asking for a minimal time and give it a weight of one. Obtain a most likely time and give it a weight of four, and obtain a maximal time and give it a weight of one. Mathematically then, you can determine what the best estimate of the time is going to be and still give weight to the usual things that might happen to delay completion of the activity or to those that will speed it up if everything should go well. If the three times were 10, 12, and 20 weeks, respectively, you would calculate as follows:

$$\frac{(1 \times 10) + (4 \times 12) + (1 \times 20)}{6 \text{ (total weights)}} = \frac{10 + 48 + 20}{6} = \frac{78}{6} = 13 \text{ weeks}$$

With that much background let's see how a very simplified problem might work. Suppose the project you were concerned with was construction of a 5-mile section of roadway into a new manufacturing plant. The activities, their priorities, and estimated times in days are:

Activity	Priority	Estimated Time
Clear and dig roadway	After survey	15
Install guard rails	After concrete poured	5
Order and receive materials	After survey	10
Paint white line	After concrete poured	1
Plant trees	After concrete poured	10
Pour concrete	After roadway ready and materials received	30
Survey best route	Must be done first	5

The next step is to prepare a PERT network showing the above. Please refer to the diagram in Figure 12-1. The descriptions of the activities were put on the diagram only to help you follow through all paths and pick the one path that represents the longest total elapsed time. It is represented by A-B-C-E-F-I-J (you may want to refer to it as AB, BC, CE, EF, FI, IJ) and has been clearly identified by a double line.

Figure 12-1.   PERT diagram of road construction.

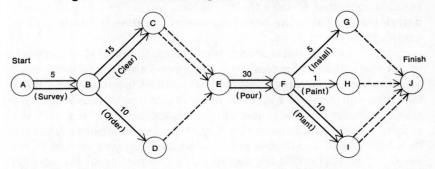

The particular path through the network from start to finish which has the greatest time would represent what is referred to as the critical path. Note that the path which takes the longest is the most critical from the standpoint of getting the job done. If the total job is to be speeded up or if it is to be merely completed on time, it is the activities on the critical path which must be most closely controlled. It would not complete the total job any sooner to expedite an activity that was not on the critical path.

A company might prepare a new PERT network each week or each month as the project progresses. As a result of running into problems at certain points and perhaps getting better than expected results at other points, it is likely that the critical path will shift from one route to another, and this could occur as often as every time the network is prepared.

The basic advantage of the PERT network is, so long as it has been accurately and completely prepared, management can always determine those activities that are of a critical nature. Thus, they will know which activities they should spend effort expediting by perhaps working overtime or incurring premium transportation costs in order to obtain certain materials quicker. With any other scheduling and controlling technique, it has been difficult to determine exactly which effort requires the most work right now, and it has been common to expend the additional effort on an item that really is not now nor ever will be of a critical nature. Do not misunderstand the meaning of the word "critical." As used here it does not mean that the others are unimportant; it merely refers to how soon an activity must be started or completed in order to finish the project on schedule.

The value of PERT is that it tends to force people to consider and provide for all the various activities of a project from the beginning to its date of completion. The technique, more than any other, also does a better job of relating each activity to every other activity. In terms of drawing the diagram, many people have found the best approach is to begin at the finish point on the project and work backwards toward the starting point. This forces you to do a better job of determining which activities must be completed before other activities can logically begin. In the event that you are personally successful in drawing the network from the beginning toward the end, that certainly would be the better approach for you to follow.

PERT can be effectively employed on a manual basis so long as there are no more than a hundred or so activities to be represented. But as soon as the volume of activities reaches an excess of that number, it is almost a necessity to obtain computer time to process the details that will provide you with information to be used for analysis. It is most likely not necessary to write your own PERT computer programs as they are readily available for most popular computers. There may or may not be a separate charge for use of the programs. An example of a computer printout of the sample problem with twenty-five days of calendar work still to be completed is shown in Figure 12-2.

PERT is certainly not to be considered a cure-all for problems relating to project control. But if it is used in a realistic manner, it should help to get the controlling job done better than any of the other techniques would. PERT is like any other technique in that it does not come free; it is going to cost a certain amount of money to put into effect and to use properly. Many companies, as a rule of thumb, will spend approximately 5 percent of total project costs on the direct efforts of scheduling and control. Within that framework, PERT is probably not more expensive than any of the other formal methods when you have considered the benefits it should provide.

PERT has a great deal of applicability in a systems project that involves many activities that are going to stretch out over a period of years. By beginning to lay out a PERT network for the

Figure 12-2.   Computer printout of PERT network.

JONES COMPANY
Status Report
Project—Route 4719
As of October 27, 1971

Activity Description	Activity Code	Percent Complete	Estimated Time	Critical Path
Survey	A-B	100	—	
Clear	B-C	100	—	
Order	B-D	100	—	
Pour	E-F	50	15	X
Install	F-G	0	5	
Paint	F-H	0	1	
Plant	F-I	0	10	X

Estimated days to completion—25

activities as previously described and by then placing an estimated time on each of the activities, it would be relatively easy to determine those activities of the most critical nature. For instance, it might be found then that the most critical time lies in obtaining and training people, or perhaps in making conversion quickly enough to begin obtaining benefits. In any event the company would be in a much better position to know at what point it would be desirable to do the best thing in terms of slowing down, speeding up, getting more people, etc., in order to meet the total overall schedule that had been established.

From the standpoint of using PERT on a systems analysis type of problem, no doubt the least precise part would be that of determining accurate time estimates for the various activities. As previously mentioned, the PERT analyst would go to the person in charge of an activity for something like building a road or a new building. Many of those activities have been performed often enough so that a fairly clear-cut pattern of the necessary time would have been established. But in systems work each system is so much different from others and because of the great variations among people who are doing the work and people whose work is being studied, it is virtually impossible to make time estimates which tend to come close to the actual time that is spent on the activity. Rather than give this as a justification for not using PERT, it might be a good reason for using it because you have a much greater problem of control here than you do in activities which tend to have better known values.

PERT, as described so far, relates to a method in which only time is controlled. Many businesses have found that they would also like to keep control of costs, and there is available a slight variation which provides for control of dollars. This method is usually referred to as PERT/COST. In this case, the amount of money required to expedite each and every activity on a per day, per week, or per month basis is shown, and management can then determine how much money is going to be required in order to expedite any activity by an additional day, week, or month. For the average company which is operating on a profit basis the PERT/COST method is probably of much more value than the plain PERT method which measures time only.

Suppose the additional costs per day that would have to be expended to expedite each of the activities in the previous example were:

Activity	Additional Costs
AB Survey	$100
BC Clear	250
BD Order	50
EF Pour	500
FG Install	100
FH Paint	Can't be expedited
FI Plant	100

If the project fell behind schedule because of bad weather or labor strikes, or if management just wanted to speed up completion of the job, what activity would they expedite? Ordering materials is certainly the cheapest in terms of additional costs, but how many days sooner would the whole job be completed if they spent an additional $100 on ordering and speeded it up by two days? The answer is zero, because that activity is not on the critical path. Surveying at $100 or planting at $100 would be the activities that could be expedited at least cost in order to speed up completion date.

The concept of PERT is often referred to as Critical Path Planning, or abbreviated as CPP.

Return on Investment

Obviously any expenditure that is involved with systems analysis and design can only be justified on the basis that there is going to be a return to the organization for the money spent. It would have to be a most unusual case where you would spend $100,000 on designing a new system if the return is only going to be a few thousand dollars a year. It must be recognized that any system that is put into effect will have a limited life in most cases because of the very nature of changing conditions, and a return that small would be considerably less than what the company could get if they merely put the money in some type of a savings account.

Within any organization there will always be various groups who are trying to obtain funds for purposes related to their specific work. For instance, there will be such groups as the:

1. Systems Department trying to obtain money for more systems projects.
2. Data Processing Department trying to obtain money to get bigger and better computers.
3. Manufacturing Department wanting to replace old equipment and put up new buildings.
4. Sales Department attempting to expand and put up more warehouses and sales offices at new locations.
5. Advertising Department so they can reach broader markets and try new media.
6. Research and Development Department asking for money so they can work on their latest ideas.

So you can see there are a lot of groups who will be asking for money, and at any time management has only so much that they are in a position to commit. Therefore, a method must be available which will permit management to make the best decisions regarding which groups are going to have funds allocated to them and how much for each of them.

Within systems work itself, you must always work on those jobs which will eventually have the greatest payoff and as a result hold off on developing programs on which the payoff is considerably less. You will be in a much better bargaining position with management if you will indicate the programs which do need work and then show what the returns might be. By using some of the techniques covered in this section you will have a much greater chance of management accepting your ideas when they see that you have followed a very logical and, as far as possible, quantitative approach in arriving at your answers. Of course there are still many managements today who do not make use of any quantitative techniques. In those cases you may not be able to use the techniques much as a selling device, but you still might be able to use them within your own department to do a better job.

One approach that may be taken with respect to this concept is that management may have set forth as one of its major policies that a particular percentage return is to be realized as a result of any investment that is made. Recognize that from an accounting sense any costs that are incurred by the Systems Department are probably not capitalized and shown on the balance sheet but would be written off as an expense incurred in each period. Regardless of the method a company uses to account for costs, you must recognize that $100,000 spent by the Systems Department must produce a reasonable return. If management has indicated that an investment in physical items should return a minimum of 20 percent on an annual basis, then the Systems Department should presumably be required to show that they can return at least 20 percent a year for systems projects.

Suppose $100,000 spent now would return $20,000 each year. Recognize that in this situation the system must continue in operation for more than five years in order to pay off. If it existed only five years, you wouldn't be getting any return at all. In fact, you would be getting less than you had put in because there was no interest added; and inflation in the meantime would give back fewer real dollars than what you had invested. So if management is really following this situation closely, they would obviously require more than 20 percent a year if the system were only expected to last for five years and have no value at the end of the period.

The concept just mentioned is often referred to as the "payback" method. It is the measure of how many years it will take to get back in the form of cash the number of dollars invested. An investment of $1 million returning $250,000 a year has a payback of four.

One of the major limitations of this method is that it does not necessarily show how many more years beyond the payback any return from the investment may be expected to continue. It also does not represent a sound way to determine which of two investments is better, since an investment with a payback of three is not necessarily better than one with five.

Figure 12-3.   Present value of $1 at a future date.

Years	10%	14%	18%	22%	26%
1	.909	.877	.847	.820	.794
2	.826	.769	.718	.672	.630
3	.751	.675	.609	.551	.500
4	.683	.592	.516	.451	.397
5	.621	.519	.437	.370	.315

In order to illustrate a better method that may be used in determining which of alternative investments is best, it is necessary to present two tables of values which would be used in this process. One table, Figure 12-3, is known as the present value of one dollar; and the second, Figure 12-4, is known as the present value of an annual inflow of one dollar. In both cases the tables have been shown in a very brief style in order to keep them simple enough to illustrate their worth. In reality the tables could be extended enough to show whatever interest rates for as many periods as you wish. For the basic purpose of illustrating the point here, the tables shown are of sufficient length.

The most important use of the table showing the present value of a dollar is this. In order to receive a dollar at a stipulated future time, you would have to invest an amount now as shown in the table at a particular interest rate. If you will look at the 10 percent column in Figure 12-3 and go down that column opposite year 1, you will find that $.909 would have to be invested at 10 percent in order to get back $1 one year from now. The interest on $.909 at 10 percent for one year would be $.091; adding the interest to the original investment would equal $1.00. Likewise, if it were your objective to get back in a lump sum $1 two years from now you would have to invest $.826 at 10 percent interest. The mechanics are as follows.

$ .826	Investment made now
.083	Interest at 10% in first year
.909	Principal at end of first year
.091	Interest at 10% in second year
$1.000	Amount returned after two years

So if your intention is to get back a lump sum of money at some future time, you would use this table to determine how many dollars you would have to invest now in order to get that return. If you want to receive a lump sum of $12,000 in five years by earning at a rate of 18 percent, you would have to invest $12,000 × .437 now, or $5244.

Figure 12-4.   Present value of annual inflow of $1.

Years	10%	14%	18%	22%	26%
1	.909	.877	.847	.820	.794
2	1.736	1.647	1.566	1.492	1.424
3	2.487	2.322	2.174	2.042	1.923
4	3.170	2.914	2.690	2.494	2.320
5	3.791	3.433	3.127	2.864	2.635

For practical reasons management will most likely require that instead of getting a lump sum return at some future time, returns will have to start flowing in at some reasonable time after the investment is made and then continue flowing in for a period of years. For this purpose you would use the table in Figure 12-4. Suppose you wanted to make an investment at 10 percent that would give back one dollar at the end of one year and also one dollar at the end of the second year. You would refer to the 10 percent column and look down the column opposite year 2. This would indicate that you need to invest $1.736 now in order to get back $1 at the end of each of the next two years. It would work out like this.

$1.736	Investment made now
.174	Interest at 10% in first year
1.910	Principal after adding interest
1.000	Return of $1 at end of first year
.910	New Principal amount
.090	Interest at 10% in second year (rounded)
1.000	Principal after adding interest
1.000	Return of $1 at end of second year
—0—	Investment value now zero

As a result of using this technique, it would be possible to roughly calculate the amount of money you would have to get back in the way of an annual return from an investment you should make in order to meet the percentage return requirement that management has established. If on the other hand you are at the point where you know what the returns will be and are trying to determine what percentage that is, you would use the values in the second table in the following manner. From the benefit and cost analysis, determine what the annual dollar return is going to be. Also determine the amount of the investment that would have to be made. Divide that return into the amount of the investment and get the dollar "factor" involved. Based upon the number of years that you would be getting that return, follow across on the line representing that number of years until you come to the column that contains the value nearest to the "factor" calculated. For instance, if after dividing the annual return into the total investment, you got a factor of 3.42 and if the investment were expected to be good for five years, you would follow across the line for year 5 and note that the column containing the number closest to it is 14 percent. This would mean that that particular investment would have a return of approximately 14 percent.

Suppose an investment of $100,000 returned $40,000 annually for four years and another of $150,000 returned $40,000 annually for five years. Which would give the better return? In the first case, the factor would be 2.5; on the fourth row of the table that would give a rate of about 22 percent. In the second case, the factor would be 3.75; on the fifth row of the table that would give a rate of about 10 percent. In deciding which of alternative investments to make, management will not always pick the one with the higher rate of return. But they are in a much better position to make a decision if this information is furnished to them.

When dividing the annual return into the investment amount, it is unlikely that you would ever get a factor that was exactly the same as any value in the table. Thus, you will have to approximate the rate; the approximation is close enough because it tells you roughly what the return will be, and that is often good enough considering the other estimates that have been made in arriving at the figures used. In the event that you are concerned about arriving at a more precise percent return, it is possible to use more detailed tables or computer techniques.

The facts available may be such that you would use the values in the table in a slightly different manner. If management has the objective of a minimum rate of return and there is an investment that will return so many dollars per year, it is easy to find out if that investment meets the rate of return criteria. Basically, you would use the table to calculate the "present value" of an investment needed. If the present value arrived at from use of the table is larger than the estimated amount of the expenditure, then financially it would be well to go ahead because you are getting the required return on a smaller actual investment. If on the other hand, the present value is smaller than the estimated amount of the investment, it would not be wise, from a financial requirement standpoint, to go ahead because you would have to invest more to get this return than you would have to if you could put the money into some other venture that would offer this rate.

Suppose management requires a return rate of 14 percent. It is estimated that an investment in a new system will be $300,000. The expected annual savings will be $82,000 for four years, after which there will be no further savings. Does this investment meet their rate of 14 percent on a present value basis? Obtain the factor from the table that corresponds to four years and 14 percent. The factor is 2.914. Multiply it times $82,000 to find the present value of money necessary. The present value is 2.914 × $82,000, or $238,948. That investment does not meet the criteria because you have to invest more than the present value to get the return.

Simulation

Simulation involves the assumption of what the results of certain acts would be without actually performing those acts. In reality you are using the technique of simulation any time you make some sort of guess about a future event as you estimate what the results would eventually be. When you are trying to calculate how much money you would actually take along for a particular day or for a trip, you are actually simulating. Typically, as a result of arriving at some particular answer, you may go through the process again using different ideas or rules in order to see what results would be on several different bases.

In business, simulation involves determining what results would be after following certain rules with specific data. Since management is held directly responsible for the outcome of a company's operations, they are the ones who should be helpful in determining the rules and should therefore become somewhat involved in any major simulation processes. The data that will be used as a base will be actual operating results taken from the most recent operating period. An exception to this would be that the future is not going to be predicated too much on anything that has happened in the past or when you are simulating any activity for which there is no data.

There may be cases where you have a very firm hold upon the nature of the data and the way it is going to act if all the factors affecting results arise within your own company's operation and control. An example would be something like labor rates over the next three years if you are in a business which usually signs a three-year contract and there is no great likelihood of wildcat strikes or conditions which would cause the company to alter that contract during the period of its normal life. In an application like inventory control, your whole technique has more of a guess nature to it because so much of the operating data comes from outside of the company. Items such as the timing of all sales orders that you receive from customers and the timing of the receipts from vendors may not be under your control as much as you would like.

Much of the written material you will find on simulation seems to deal in complicated or tedious mathematics. But it is possible to do a reasonable job of simulation without getting that deeply involved. Depending upon how deeply you care to get into it and on how many items you wish to simulate, you might wish to employ methods which are far beyond the scope of this book. Examples of how simulation may be used are as follows.

1. What would total payroll costs have been last year if certain labor rates and certain fringe benefit rates had been at higher levels? A number of companies use the simulation technique at bargaining time when the labor union comes in with its demands. Knowing how that particular payroll worked in the past period with so many hours of regular work, hours of overtime work, and so many people on the payroll, it is relatively easy to run through last year's actual time worked using the projected rates to see what last year's costs might have been. At that point, management is in a much better position to determine how well they might be able to afford the requested rates because they could relate the answer to what next year's payroll costs might be.

2. How much should a business be capitalized when it starts operating, realizing that such things as accounts payable and accrued items if allowed to build up and go for a certain period of time unpaid would have the same benefit to the company as a greater investment on the part of the owners? At the same time

the people about to go into business might want to simulate what the cash flow would be, based upon a policy of selling for cash only or perhaps on a limited or very extended credit basis. Depending upon company policies with regard to sales terms, it may be possible to buy something on credit and sell it for cash and then actually use the cash from the sale to pay off the vendor. Various approaches may be used to eventually show what the optimum capital investment would be.

3. At what future time will cash run out and have to be replenished under various operating policies of paying off short and long term debts, paying dividends, investing in certain fixed assets, or perhaps putting more money into a research program?

4. What is the best method to finance expansion considering that money could come from common stock, preferred stock, or from long-term debt?

Simulation is typically done in such a way that you try a set of data with certain rules and then try different data with those same rules. Then you hold the data constant and apply different rules. You continue these processes until you find results that appeal to you. Of course, it takes a great deal of management intuition to know the point at which something is really good or not good. As has been pointed out on several occasions, a business must be willing to look at various methods of operation in order to eventually select the most appropriate one.

Figure 12-5 shows how simulation might be used on an inventory item to determine the behavior of inventory levels using different reorder points. Note that orders on 2/6 and 2/9 could not have been filled immediately if a reorder point of 75 had been in effect on this item.

There is one popular use of simulation in which the outcome is not going to be changed based upon whatever method has been applied. That is the task of weather forecasting, where the people who apply the technique obviously have no control over the results which are going to occur. Of course, the future may hold the possibility that man may be able to control the weather by applying certain methods to obtain the type of weather that is desired.

Sampling

Sampling is a systematic procedure in which you observe only a portion of the transactions which occur. Based upon the nature

Figure 12-5.   Simulation of inventory levels.

	Actual Using Reorder Point of 100		Simulated Using Reorder Point of 125		Simulated Using Reorder Point of 75	
	Activity	on Hand	Activity	on Hand	Activity	on Hand
1/1		129		129		129
1/2	− 12	117	− 12	117	− 12	117
1/5	− 15	102	− 15	102	− 15	102
1/12			+250	352		
1/18	− 24	78	− 24	328	− 24	78
1/28	+250	328				
2/6	−141	187	−141	187	−141	(63)
2/9	− 12	175	− 12	175	− 12	(75)
2/16					+250	175
2/24	− 60	115	− 60	115	− 60	115
3/3			+250	365		
3/4	− 42	73	− 42	323	− 42	73
3/14	+250	323			+250	323
3/15	− 62	261	− 62	261	− 62	261

Reorder quantity of 250; average 10 days to replenish.
+ = Receipts
− = Sales

of the techniques used and the data studied, some conclusions are drawn regarding what all the data must be like.

From a business information standpoint, samples are usually taken for two purposes. First, results can be obtained faster, and second, the cost of obtaining those results can be minimized. For instance, the TV ratings that we hear so much of in connection with audience preference and cancellations of TV programs are based upon a small sample of viewers. A few thousand TV sets located throughout the country are automatically monitored on a continuous basis throughout viewing hours. The information collected is that of whether the set is on or not, and if on, what channel is being viewed. If 27 percent of the sets monitored are turned to a particular program, then 27 percent of all people are presumed to be watching that program.

Another interesting sampling system is used on a great volume of data within the airline industry. If you were to fly from Portland, Maine, to Pittsburgh, Pennsylvania, you would need to make at least two legs on two different airlines. At some time after you had completed the flight, the airlines involved would split the money you had paid for your fare. (No doubt you would have bought a through ticket and paid the first airline used on your trip.)

Consider that there are thousands of trips like this every day. It thus becomes an enormous job for the airlines to split the money according to the ratio of the trip that each airline served. Instead of figuring the exact split on each flight, many airlines have agreed to use a sampling technique where they test representative flights during the month and then split interline fares on that basis.

The sample permits the job to be performed much quicker and at a considerably lower cost than would be the case if each interline fare were separately accounted for.

The use of sampling should be considered in the study of a present system and also in the design of the operation of any new system. Even where it might appear at first glance that sampling would not apply, significant results might be obtained by putting it to use on a logical basis.

In analyzing a present system, some form of sampling might be used in a fact gathering scheme as follows.

1. You usually need to know something about the number of transactions there are so you can properly provide for people and equipment needs. Instead of counting every individual transaction, you may count only several day's worth to project the total for a month. Or you may study only two or three months in order to estimate the annual volume.

2. If it is an error rate you are looking for, you might look at every fifteenth transaction or so and count them. Then you can project what the total situation must be like.

When you have reached the point of designing steps in a new system, there are many ways of using sampling techniques.

1. The cycle at which a physical inventory is taken may be stretched out. In companies that have well-designed and well-operated inventory systems, a full physical inventory need not be taken each year. If it can be shown on a sample of items that the quantity on hand is equal to accounting records, then it is generally assumed that this is true in the case of all items. This not only saves the time and expense of counting and processing so much data, but also possibly eliminates temporary shutdown of the warehouse area as a full count so often involves.

2. The preparation of volumes of paperwork can perhaps be eliminated. Suppose a system has to have steps to check the quality of items coming off an assembly line. In the same way that a photographic flash bulb manufacturer can check only a sample of his product, most other manufacturers can develop sampling techniques to check only a few of their products. Less inspection means less paperwork to be prepared, processed, and stored. But systems analysts must work very closely with manufacturing people on a point like this.

There are two very simple methods that can be used to choose the items that will be studied. One uses a table of random numbers. This is merely a listing of numbers in which each value has just as much chance of appearing as any other; there is no pattern to their listing. Suppose you have ten people whose work you want to sample on a random basis. You might find that a random number table in a statistics book lists the numbers up through 10 as 2, 9, 3, 7, 10, 1, 1, 8, 4, 6, etc. By informally assigning each of the people one of the digits up through 10, you could pick them out for review according to the way they were listed in the random number table. Most statistics and auditing books contain extensive lists of random numbers.

Another approach works like this. Suppose you have decided you want to check about ten percent of the transactions that occur in a department. If there are normally about a hundred in a given period of time, then you would want to check every tenth one. You might pick out the fifth one, the fifteenth, the twenty-fifth, etc.

There are times when sampling might not be appropriate. If you were at the point in a systems study where you were collecting copies of the various forms used and reports issued, getting only a sample of them might not be enough to help you determine what the total picture was like. If you were trying to determine system costs, you probably couldn't do it based upon a sample of actual department costs. The reason for this is that the number of departments is probably small. It would be difficult to pick a sample that would be representative of all of them.

Sampling can be mishandled to the point of becoming worthless. The manager of a department of clerical people decided to use a sample basis to find out what his employees spent their time doing during the day. He prepared a form with the following format.

EMPLOYEE	DATE					
John Doe	*4/15/71*					
ACTIVITY	Time Observed					
	9:20	10:04	11:37	1:45	2:02	
1. Working at desk, alone	X					
2. At own desk, with co-worker		X			X	
3. At desk of co-worker						
4. Talking on telephone			X	X		
5. Not in department						

He advised all his employees ahead of time that he would be circulating among them at random times during the day to record what they were doing. The study was to last for three months, after which he would try to develop something so that people could make better use of their time.

When the program got underway, at random times during the day he would record what the status of each employee was. But the employees had not been convinced as to the usefulness of the whole approach. When they saw him coming with his forms, they intentionally switched what they were doing. The results he collected were so distorted they were meaningless. He soon realized this and abandoned the program. It had never occurred to the manager that he could have gotten accurate results by staying at his desk and collecting the data in a more informal way.

From the brief discussion above, you can see that a good sampling program can be very effective in a systems environment. What is needed to put it to good use is a knowledge of the operation and some creativity on the part of the systems analyst. The modern auditor lives within a concept of sampling through all his work. If systems personnel will cooperate with auditors in the design of systems, a lot of value can be obtained from such programs.

Work Simplification

Work simplification is a technique that should be used extensively in any systems work in order to ensure that every step is performed in the best possible way. The best possible way is often the simplest one. This does not necessarily refer to the process of setting the objectives of the program but to the manner in which specific steps are performed. It is meant to concentrate upon the details of operations.

The basic idea behind this technique is to find the most efficient way to get a job done. Obviously, you should be employing the method as best you can during the design of a system. Since you will never be able to design it completely in the best way and because of changing conditions, you should also be going back later on a follow-up basis to try to simplify whatever possible.

The major ways in which work simplification can be accomplished is through the use of these processes.

1. Try to eliminate unnecessary forms. A copy of a report which is merely filed can perhaps be eliminated.
2. Eliminate those steps which perhaps represent a duplication of effort. A company eliminated a report which recapped orders received because the basic information was available from a report which recapped shipments.
3. Combine steps. Instead of having data go to the same work station at two different times, have everything completed the first time it arrives at that station. Recall that in Chapter 7 one pass of sales orders against the file was eliminated.
4. Rearrange steps so that they may have a more logical flow.
5. Instill in others a plain method of simplifying things and making them easier to perform. Car manufacturers have made all car prices in whole dollars, and airlines have made all fares in whole dollars. This has made tax calculation easier and has reduced errors.

Another technique that should be performed in conjunction with work simplification is the one referred to as "office layout." By using reasonable techniques here in making sure that work flows back and forth through the office in a most logical manner and on a path that best represents the shortest distance to the next step, the entire procedure will be performed easier and perhaps with less expense. People will thus feel that you really care about the things that they have to do in order to perform their jobs.

Work Measurement

Work measurement is a technique that is used to determine what represents a reasonable amount of time and cost in order to perform a particular step. Many attempts at setting up work measurement programs have resulted in very unpopular situations because people get the feeling that management is breathing down their necks. Because the analyst may have gone into a department and used a stopwatch to time people or may have taken movies of the department in action, workers have gotten the idea that such programs were part of a campaign to speed them up. But we must recognize that the typical labor force has been subjected to some version of work measurement over a great period of time, and it is only natural that programs of this type also move into the business systems area. The method used to approach the problem will have a great bearing upon how people feel about it and its success.

In order to apply the technique, a work measurement analyst might break down a particular job into its smallest components. He might calculate the amount of time it should take for a clerk to move her hand from one side of her desk to the other in order to accomplish a necessary step. He would eventually add all other components together and arrive at a standard time for a particular transaction to be processed. By relating that amount of time to a unit such as an hour or a day, it is then possible to determine how many transactions a day a particular work station should be able to perform. Quite often it is found that the amount of output which can reasonably be expected is considerably higher than the present output. Perhaps good methods of work simplification may never have been applied; perhaps department supervision never did anything about this point or may have never properly trained the people.

One company applied work measurement in the following way. An analyst knew that typists hired into a particular clerical level had to be able to type sixty words per minute. Occasional observation of such typists revealed that many of them did not seem to be under undue strain to obtain that speed. The analyst also knew that girls in the Keypunch Department seemed to be even more casual in their work, and they surely must be capable of a greater rate than what they were producing.

He knew the Data Processing Manager expected 10,000 key strokes per hour from a trained person. So he decided to calculate the equivalent key strokes per hour of a typist.

60 words/minute × 5 characters/word × 60 minutes/hour = 18,000 key strokes/hour

The analyst immediately saw that many key punch operators could go much faster if there were only some incentive to do so. He proposed to management that an incentive pay scale be adopted so that key punch operators who attained a speed in excess of 10,000 key strokes per hour would be paid an excess amount provided their accuracy rate was maintained at a reasonable level. The scale was set up so that both the company and the employees benefited. If he had made the proposal only on the basis of getting the girls to work faster, it may not have been accepted at all. It did cost money to administer the program over and above increased salary costs, but it was well worth it.

Equipment Selection

There are various approaches that might be used to select the equipment that will help people operate a system. One way is to take the easy approach and let an equipment manufacturer make the feasibility and systems studies and then allow him to select from his own line of products the equipment configuration. Despite the inadequacy of this particular method, it has been used by many companies for years. One of the drawbacks is that a representative of a computer manufacturer will most always find his own equipment serves the purpose well, and he may be able to sell more to the user than what the user actually needs.

Another means of arriving at the brand name of equipment to be used is through what may be termed "reciprocity." In this situation, two companies may tend to buy products and services from each other. For instance, suppose that an equipment manufacturer ships all its products over the lines of a particular freight carrier. When the carrier gets to the point of buying equipment, he may decide to buy from that vendor because it will enhance his relationship with the vendor and assure him that his transportation system will continue to be used. Another example is a bank that may be holding substantial deposits for a manufacturer and at purchase time may decide to buy from that manufacturer in order to continue the deposit relationship.

The federal government often goes through a very lengthy process to select an equipment vendor. It may follow the policy of choosing on the basis of the lowest price, the smallest vendor, or the vendor that happens to be doing his production work in a geographical area that may have a depressed labor market. Another approach that some companies have obviously chosen is to stay with a particular vendor regardless of what other vendors may have available or may be offering in the way of incentives.

The foregoing examples represent approaches that may not take into account many of the practical aspects of procuring and working with a particular category of equipment. However, such methods are widely used and cannot be faulted too much as long as the buyer recognizes what he is getting and that perhaps it is not the best procurement job that might have been done. Many users would rather apply as objective of an approach as possible in order to arrive at the best buy for their particular situation. The remainder of this section will try to relate to objective ways of arriving at equipment selection.

An objective method certainly does not involve looking through catalogs or calling in vendors or visiting other people to see what equipment is available and then trying to adapt that equipment to the operation. An organization must take a completely opposite approach and determine what it is that they want done and then go out and obtain equipment that will serve these needs. After determining exactly what it is that you want equipment to do, you should go through an analysis such as follows in order to pick the exact configuration and vendor.

1. What type of self-checking features does the equipment have? That is, can it detect any errors that it has made itself?
2. Does it have any programmable method of checking on errors in data that has been entered into the machine?
3. How can errors be corrected once they have been recognized?

4. What is the speed of the machines? Preferably the speed should have meaning only in terms of the number of transactions the machine can handle in a period of time as opposed to any quoted speeds in microseconds, nanoseconds, lines per minute, etc. It is not the rated speed that is so important, but the "throughput" speed.

5. How much help can you realistically expect from the manufacturer? Does he provide educational services that will teach the proper use of the equipment? What kind of help will he provide, either on a paying or nonpaying basis, in terms of helping set the device up and getting it to work?

6. Are there other users who have experience with this particular device, or are you going to be a pioneer?

7. What problems are involved in physically preparing the location where the machine will be installed? Does it require air conditioning and humidity control and are there any special power requirements? What are the space and weight limitations?

8. How available and expensive is it going to be to get accessory items that go along with the device?

9. What problems are going to be involved in converting from the present method of operation to the method which employs this device? This phase of operation can be very trying.

10. Where can a like piece of equipment be seen in action? It is important to spend a reasonable amount of time on this so you can better see what operating conditions are like. Hopefully you can ask questions there and see what others' experience has been.

11. Are programs and appropriate aids available and working? This particular point is one on which much equipment has been ordered on the faith that software packages would be available and working at a certain future date. But many user companies have found that delays up to a year or so are fairly common and some of the programs don't even work when they do get them. There was a time in computer history when all such programs were free and many users had the tendency to overlook the fact that they were late or did not work well. With more programs of this type having a charge attached, the user is more

likely to expect them to work as soon as he starts paying for them.

12. It is desirable to obtain a modular approach. This refers to the fact that additional components or more storage can be added to the existing machine. Some manufacturing processes are set up so that any additional devices would require that present equipment be returned and then be replaced by a completely new unit that has the additional uses or capacity.

13. Specifically what guarantee is being received with the device? Exactly what do you get for the price you pay? In the event that you decide to return the device, is there any remote possibility that you will be able to do this without penalty?

14. Some companies prefer to obtain various portions of their equipment from different vendors so that they can adequately compare operating conditions among the different vendors. This can be used as a wedge to obtain concessions from all involved if the sources can see competition coming into play.

15. The equipment buyer must get into a position eventually where he can determine the real difference, if any, between competing products. Many people today feel that a device of one company is not a great deal different from a similar device that most any other vendor has and that any differences at all are going to be related to the services that the vendor can provide.

16. Some companies buying computers want only the equipment and not any of the other services. Some of these users are so large they have established all the training, programming, and repair services that may be needed and thus, want only to buy and pay for the hardware itself.

Obviously no vendor is going to be able to score a 100 percent rating on all these questions and points that have just been listed. As a result, many companies have found it convenient to determine what portion of their total selection criteria will be based upon the major items previously reported. After preparing an analysis such as that which follows, a company can use a somewhat objective approach in order to arrive at a specific piece of equipment and the vendor who is going to supply it.

Criteria	Maximum Value	Vendor			
		A	B	C	D
1. Self-checking features	5	5	4	4	5
2. Average throughput speed	5	5	2	4	4
3. Manufacturer support	10	0	8	5	0
4. How long machine has been in service	15	15	0	10	8
5. Installation costs	10	5	2	5	8
6. Conversion problems	15	10	3	8	12
7. Ease of programming	30	25	10	20	18
8. Cost	10	3	10	5	5
	100	68	39	61	60
		Vendor chosen			

Another way to perform a similar analysis is as follows.

Criteria	Vendor Ranking			
	A	B	C	D
1. Self-checking features	4	2	2	4
2. Average throughput speed	4	1	3	3
3. Manufacturer support	2	4	3	2
4. How long machine has been in service	4	1	3	2
5. Installation costs	3	1	3	4
6. Conversion problems	3	1	2	4
7. Ease of programming	4	1	3	2
8. Cost	1	4	3	3
	25	15	22	24
	Vendor chosen			

In the second approach, a vendor is given a rating of four for best, three for next to best, etc. On this basis, vendor A still gets the best rating, but D now moves ahead of C. With either rating scale, the difficult thing is gathering enough data to determine who has an edge on each of the factors.

Questions

1. Why is so much data transmission required, both with regard to source transactions and completed reports?

2. What basically determines whether physical or electronic means are used to move data?

3. What enables computers to "talk to each other?" What is actually happening when this occurs?

4. What is the name of the transmission method which can send an exact image of a printed document? Name several instances where it would serve a useful purpose.

5. In PERT, what is the difference between an activity and an event? How are they related?

6. Activity lines on a PERT network must be drawn in proportion to their respective times to complete. True or false? Why?

7. Why is it desirable to consider optimistic, pessimistic, and most likely times when preparing a PERT network? When would you just use the most likely time?

8. Specifically, why might a company be willing to spend a little extra money to get a project completed early? How might PERT help accomplish that?

9. What is the difference between PERT and PERT/COST?

10. What is the payback on an investment of $400,000 which returns $20,000 a year? Can a business afford this investment?

11. It is expected that a systems change will require an investment of $50,000 and return $12,000 annually for seven years. How many estimates are used in calculating the percentage return? Make a brief list of things which might happen to throw each of those estimates off by a substantial amount. To how many decimal points should you carry the percentage return? Why?

12. What would you consider to be the soundest ingredient needed for an effective simulation process? Justify your answer.

13. Why is a computer so well-suited for simulation?

14. How can you sell a work measurement program to employees so they don't think it is some kind of speed-up campaign?

15. What is reciprocity? Can its use be justified? Why?

16. Give an example of how the lowest bid, if accepted, may turn out to be the "highest cost."

17. From the activities shown below, sketch a PERT network and
clearly illustrate the critical path.

Activity	Most Likely Time
Brief seminars for officers of company	10
Classes to train operating employees	30
Complete conversion to computer	5
Computer installed and plugged in	5
Conventional punched-card equipment returned to manufacturer	3
Parallel operation	40
Problem definition	30
Receipt of new cards, paper forms, and magnetic tape	10
Selection and ordering of specific computer	10
Systems and programming	100

Other Readings

Arkin, Herbert, *Handbook of Sampling for Auditing and Accounting*. New York: McGraw-Hill Book Company, Inc., 1963. This book
is a very easy-to-read, nontechnical presentation of the uses of
sampling. Appendices contain those tables of numbers useful in a
sampling program.

Data Communications Glossary. Form C20-1666. International Business Machines Corporation, 112 East Post Road, White Plains, N.Y.
10601.

Data Communications in Business: An Introduction, American
Telephone & Telegraph Company, New York, 1965. A good introduction to methods of communicating data over telephone lines.

Horngren, Charles T., *Accounting for Management Control*, 2nd
Edition. Englewood Cliffs, N.J.: Prentice-Hall, Inc., 1970. Chapter
14, along with its appendices, describes various methods of determining return on investment.

Martino, R. L., *Finding the Critical Path*. American Management
Association, Inc., New York, 1964. A very readable coverage of the
PERT technique.

Neumaier, Richard, "A Look At Work Simplification," *Ideas for Management,* 1964. Gives very worthwhile guidelines for setting up a work simplification program along with 21 specific proposals put into effect at one of Mr. Neumaier's clients.

Office Clerical Time Standards. Association for Systems Management, Cleveland, Ohio, 1960. Suggests reasonable standards for dozens of common activities associated with office procedures.

CHAPTER 13

Barriers and Pitfalls

Most of the stumbling blocks to the design and implementation of effective systems will result from necessary dealings with people. As one president of a company said near the end of a lengthy project of constructing a new office site, "If we would have only been concerned with steel, bricks, and mortar, things would have gone well; but since we had to deal with people at every turn of the road, we had problems."

Lack of Solid Objectives

A prime reason for lack of success in any endeavor is the failure to insist upon and/or provide solid objectives from the very beginning. How many times have you heard of the following situation? A person has been given a certain job to do, with only a brief description regarding any of the specific details. He performs the task as he understands it and shows it to the requestor who says, "It's okay, but it's not quite what I had in mind." Of course, it's not quite what he had in mind because he never told the worker enough about what he really wanted. Perhaps he didn't have any good suggestions himself when he gave the assignment, but once he has seen something done, he at least knows more about what it shouldn't look like. Good, adequate systems design is not very likely to result by accident; it will come about only as a result of thoughtful, hard work by many people who cooperate well.

When an organization has decided to buy or build a physical structure, they generally have a pretty good idea how it is going to be used and what kind of a payoff it is going to have. There is no good reason why they should not apply the same practice to the expenditure for a nonphysical item such as an information system.

It will not suffice to just tell the designers to make the new system better than the old or to make it better than that of the competitors. It might take only a 1 percent change over the present system to comply with that standard.

I have seen instances where management has requested employees at various levels to prepare a list of their personal objectives. This has been done so that management can learn employee interests and help them determine where they should be headed. Presumably management will do what it can to help the employees reach those goals. Management must also set its own objectives so employees know what they have to do in order for the company to move as it should. Requiring a ten percent increase in profits every year is a much sounder objective than to freely state that everyone should do the best job he can.

Failure to Communicate

The sales manager of a manufacturing company was visiting one of his customers to find out how their new machine was working. The customer was pleased with the machine, but their production manager wanted to know how they could slightly bypass one of its safety features in order to get greater production. The sales manager declined any knowledge of how it might be done. He was later describing the incident to a lawyer friend who told him they should immediately send a registered letter to the customer informing them that they should not make any alterations to the machine in the interest of safety. The sales manager shrugged the suggestion off with the statement, "That is a matter for the operations people." Two months later the daily newspaper carried a front-page story telling how a worker at the customer site had been badly injured using the machine, and the customer immediately sued the manufacturer; it seems the customer had a different recollection as to the salesman's advice regarding use of the machine.

In another situation, a company had a very high error rate in the recording of inventory transactions. Systems and operations management shrugged off the problem with statements like "That's all you can expect from help these days." Over a year after the problem had gotten progressively worse a systems analyst casually asked an inventory clerk what was wrong. He was told the form that was being used for recording was too complicated. Operating people had never been asked before, and they had not volunteered the information.

While poor communications may be the result of a person's inability to express himself well verbally or in writing, a great deal results from assuming that others know what is going on or believing they don't need to know. Some of this problem can be overcome by educating people as to how their positions relate to other positions in the organization.

Company Policy

Company policy can impede good systems development. Policies can be improperly interpreted or allowed to continue long after they have served their intended usefulness. Several years ago I accepted the sponsorship of a college sports car club. Despite all of our very precise flyers, advertisements, and articles that all driving events were to be of a non-speed, non-horsepower, any-kind-of-car-allowed type, there were students who said they didn't come to events because they knew they wouldn't be able to compete with the Corvettes. So the organization dropped the name "Sports Car Club" and adopted the name "Auto Club," to show that everyone was welcome. But it didn't completely overcome the image many people had of its being a high-powered car organization. Here are some examples of unusual interpretations of policies.

1. One state has a law stating that all businesses selling food and beverages must provide restroom facilities. A ten-year-old boy selling lemonade on a street corner was put out of business until the governor interceded and got the appropriate bureau back to work.
2. The sales tax office of another state so literally interpreted the tax laws they tried to force children selling lemonade to collect the sales tax.
3. One large city pays its police force on Wednesday, the pay to cover work up through Saturday of that week. When a policeman was killed in the line of duty on a Thursday, the city auditor tried to force the policeman's widow to return the excess payment to the city.

Policies are normally established by top management to guide employees in making their own decisions rather than going to management every time a slightly different situation develops. But each policy surely was never intended to last forever or to stand in the way of permitting business to improve itself. For instance, a company may have a policy that says any customer's account that is not paid off by a certain day will be subject to a service charge. But if it is obvious that payment was not received because the area was practically immobilized at that time by a blizzard or a flood, the policy should be relaxed; the system should be designed flexibly enough so that the service charge won't be imposed. Any income that would be earned due to imposition of the service charge will not weigh as much as the ill-will gathered by following the rules. So often in a case like this, the existence of the policy is given as justification, but that is merely an excuse for an inflexible system. Management may know so little about the design of the system they can't point to anything other than the policy.

In a bank in California the analysts set up a system on the assumption that the bank would never make errors of the type that would require adding a deducted amount to a customer's checking account. When a valuable customer proved that the bank had deducted from his account in error, the bank found that the only way to rectify the mistake was for their Public Relations Department to treat the customer and his wife to dinner. It is doubtful that the bank management ever had this type of policy in mind.

Savings and Benefits Hard to Pin Down

Despite the amount of planning which may have been done, it is very difficult to closely estimate what the dollar savings or net benefits of a new system will be. Even when the new system is finally in operation, it will be hard to get a clear comparison between new and previous operations because so many conditions have changed in the meantime.

This doesn't mean that most practicing systems analysts won't make a savings prediction while justifying their position. Some have been known to carry all financial projections out to the penny. They can readily report what will happen and place a dollar tag on the various components. But there will always be some differences of opinion as to what will happen between what the analyst and operating people think, and these differences must be resolved to a reasonable extent under the work of Chapter 8.

Recognize that it may be at least several months to a period of years from the time the study was made until installation is complete. Absolute comparisons then become difficult to make for the following reasons.

1. Volume has probably changed substantially. Many companies show an annual growth rate of 10 to 20 percent or more.
2. There may have been considerable changes in the nature of the work. Perhaps some products were dropped; new ones may have been added; work is being performed in different geographical areas; there may be changes in the tax laws or union requirements.
3. A change in top management personnel has considerably altered many things. Perhaps this has even changed some basic company policies.

4. Some things of a very important nature were just overlooked at the time of the study. One person who designed a brief system that would be used to print magazine mailing labels found that the forms he had bought were so narrow they would not fit into the printer on his computer. His work had temporarily no value because the system wouldn't work, and the benefits he expected were partially absorbed by the need to buy more expensive forms.

However, from an overall standpoint, the important consideration is making sure that objectives of the system have been reasonably met. If they have, concern about the details of just a few points is not necessary.

Resistance to Change

This is basically related to fear on the part of people that they personally will suffer in some way if a certain change is made. From a practical standpoint the change, when put into effect, occasionally does cause difficulty.

One large company was able to eliminate forty secretarial jobs when it took the office secretary away from certain management levels and installed a centralized dictation center which users tapped by telephone. A manager was assured of the same service under the new system, but even recognizing the sizeable cost savings that would come about was not enough to make people happy with the change. It was a plain case of many hurt feelings because many managers no longer had what they considered to be a "private secretary."

It is difficult to get people to immediately cooperate just because some management action has been taken. For instance, when the New York Central and the Pennsylvania Railroad merged into the Penn Central Railroad in 1968, management found that people in the merged offices did not immediately cooperate. The railroads had been arch business rivals for over one hundred years, and the merger could not immediately affect negative employee opinion.

In one company the major deterrent to placing everybody on a weekly pay basis was that some people despised being paid on the same day of the week as production people. Apparently this was one of the last status symbols the salary people had.

Resistance to change cannot be completely overcome, but its effects can be reduced by the following.

1. *Education.* Clearly show that the necessary change will benefit the business and hence its employees.
2. *Reporting results.* Keep a reasonable log of previous changes and the positive results that came about. In those cases where there were failures, don't cover them up, but show why they happened.

People Try to Beat the System

Anyone who is designing a system must keep in mind that there are always going to be people who will try to "beat the system." A number of years ago most airlines put in a one-half fare for persons between twelve and twenty-one years of age. The person would merely be on a standby at the airport and if there were seats available a few minutes before flight time, the airline would allow those people to fly at one-half the regular fare. The airlines were willing to do so because they were happy to accept a one-half fare rather than having a seat go completely empty. Many people immediately found that a way to beat this system was to make fictitious reservations for others, and then when those people obviously did not show up they themselves were available to take the flight.

In December 1969, a tobacco company in Canada had to withdraw a promotional stunt they had been using. There were codes on cigarette packs that because of certain winning combinations could be turned back to the company for cash. Some clever people determined a way of spotting the winning packs, and one person bought 4000 packs of cigarettes and turned them in for $35,000. And this man was not even a smoker! It is estimated that the company paid out in excess of one quarter of a million dollars, most of which went to people who had been able to beat the system.

Role of Factors from Outside the Business

A business will always have enough problems trying to control all the people, equipment, and processes under its own roof. But it can really develop additional troubles because of the role of factors of outsiders, those over which they may have little or perhaps absolutely no control. This sample list shows such things.

1. A fire, flood, or storm may physically destroy facilities. Even if insurance does cover most of the financial loss, the reconstruction period can be very agonizing. In 1953, fire destroyed a General Motors transmission plant. Even though the loss was covered by insurance, General Motors had to temporarily put transmissions into cars for which they were not suited. A customer relations problem developed.

2. Breakdown in service by a monopoly operation such as the U.S. Post Office, telephone company, or electric company can occur.

(a) In 1967, service at the Chicago Post Office completely collapsed, and it took weeks to get mail flowing smoothly again. In 1970, the first mail carriers strike in history left many areas without mail service for several days. Many companies had to borrow money to meet their payrolls because they were not receiving customer payments through the mail.

(b) Certain cities have had serious interruptions in telephone service because of substantial increases in demand and the inability of the companies to expand quickly enough. The telephone company in New York City borrowed thousands of equipment installers from affiliated companies throughout the country. In 1970, one telephone company had as an immediate objective that every customer get a dial tone within ten seconds after picking up the phone.

(c) On November 8, 1965, much of the eastern seaboard and its 30,000,000 people were without any electrical power for about four hours because of a power failure. On extremely hot days, some electric companies have had to ration electrical power because of abnormal use due to air conditioners; some businesses have had to close and send their employees home.

3. A vendor may go on strike, may go out of business, or may not be able to allocate any more of a scarce commodity to you. He can also cause you considerable trouble by merely missing a deadline.

4. You may have been selling a great percentage of output to a single customer, and that market may have disappeared through no fault of your own. In one company total employment went from 1200 to 300 in a two-month period because of cancellation of a government contract. That cancellation even eliminated the Systems Department and all other areas that were related to improve-

ments and cost cutting. In another situation, Company B said they would gladly buy from Company A if the appropriate invoicing were done on Company B's forms. A problem arose at Company A because they were using a computerized invoicing method that could hardly be altered for this one exception. They resolved the problem by doing that one customer's invoicing manually in order to get the business and satisfy the unusual requirement.

5. Unplanned acts by employees cause interruptions. Resignations and early retirements can cause disruptions in systems operation. A new and highly recommended employee may not turn out to be nearly as good as had been planned. A supervisor who seemed to be doing a good job may have changed his tactics and results. The story has been told of two people who had both just severed employment with their respective firms. Mr. A said he had caused problems to his previous employer on the last day of work because he threw away a deck of cards that represented one whole computer program. Mr. B said he had just removed one card from each of 800 programs at his previous company.

6. There are other significant causes.

(a) Because of the Mid-East War in June 1967, the Suez Canal was closed to all traffic. Ships in the Canal at the time were marooned. Shipping firms that had been using the Canal had to drastically alter their routes.

(b) An airplane crash completely destroyed thousands of checks enroute from the banking system back to the drawing company. The firm had to go through detailed procedures to find exactly which checks they were.

(c) You must be very cautious of any machine demonstrations you see. Several years ago I attended a computer demonstration of one of the major manufacturers. The demonstration was set up where several hundred people could see the machine operating on the stage of a large theater. The computer had card input, card output, printed output from both the line printer and typewriter, six or eight tape drives, and one disk file. One of the strong points of the sales pitch was the fact that these devices could operate simultaneously. During the demonstration the computer operator was kept busy feeding cards into the card reader, removing cards from the reader stacker, placing cards into the punch, and removing them from the punch stacker. This operation kept him busy practically all the time, and the other components required no action on his part. When the formal part of the demonstration had been completed, the audience was allowed to walk around the machine to notice its physical features. I observed there was only one box of 2000 punched cards behind the card reader/punch. Further observation revealed that the cards had no holes in them. I asked the computer operator what kind of a demonstration that was and he mumbled something to the effect that they could have had holes in them but it didn't matter in this particular case. I also observed that the printing that was being done was just one line of data from core storage that was merely being printed in a repetitive fashion. Obviously, the input and output devices had no true relationship to any real job that might be performed. It would have been a mistake for most companies to get that computer model at that time because the manufacturer clearly had not developed programs adequate for use.

(d) Several years ago a person who had been in my data processing class three months told me of an experience of his as a night-time computer peripheral equipment operator. Two or three times a week he was assigned to a long punched-card sorting job. There were 30,000 to 40,000 cards to begin with. He sorted on column 12, kept the cards from stacker 5, and threw all others into the trash. The retained cards were sorted on column 22. Those in stacker 7 were saved and all the others discarded. This pattern was used for a total of seven sorts. The final number of cards saved was usually less than a dozen. After he had done the job several times he studied what was happening and found he could sort in a much different order and get the job done in *one-fourth* the time. He chose to use the revised way the next time. The department supervisor saw something different was taking place and asked him about it. Despite a good explanation, the supervisor said he was to revert to the way he had been trained, that the job had been done the other way ever since it had existed, and furthermore the worker "wasn't being paid to think." The purpose of pointing this out is to show that the whole team is not always pulling in the same direction; that deficiencies that are so obvious they come to the attention of the management of the operating departments are not always going to be solved there and you may only run across them by accident; that a competent specific group such as a Systems Department must be constantly looking for ways of improving.

(e) In the automobile industry certain laid-off workers get 95 percent of their regular pay. At Christmas time in 1969 the workers at one automobile plant demanded to be laid off for two weeks just as most others in their industry were. Apparently they felt that a 5 percent reduction in gross pay was a very small price to pay for a two-week paid vacation at that time of the year. (Note that the employer would not be getting any return for those costs.)

(f) Some companies have had the problem under quick response systems that reports often arrive at top and middle management desks at the same time. Many middle managers have complained about such promptness because they can no longer cover up when the supervisor calls immediately and questions them. In the past, the middle manager had enough time to justify results or perhaps alter reports as he desired to send on to top management.

Questions

1. Shown below are some general statements about company objectives. Restate them into positive, attainable ones:
   a. We will try to increase our profit as much as we can.
   b. In order to prevent major problems, we will limit sales to any one customer to a reasonable amount.
   c. We will make our best effort to pay bills when they are due.
   d. One of our personnel policies is to provide the best life insurance plan available.

2. List some of the things which make "communication" such a big problem.

3. What is meant by the term "company policy"? Who sets the policies? Why? Are they expected to exist for the life of the company? What should you as an individual do if you find a policy that seems to be costing the company an undue amount of money?

4. How reasonable would it be to round all cost and savings summaries to thousands of dollars? Why?

5. Why is it so difficult to accurately estimate cost savings?

6. A company met all its stated objectives but their new system was really a failure. How could this be?

7. Do people basically resist change or do they just resist those changes which they fear? Explain.

8. Why has there been such a tendency for companies in similar industries to merge? What types of systems problems may such merged companies have at the very beginning?

9. Is it possible to obtain the proper insurance to cover against any mishaps that might occur? Support your answer.

10. What could happen to a vendor of yours that might cause serious trouble in the operation of your system?

CHAPTER 14

Other Comments About the Systems Field

The preceding chapters of this book have attempted to show the general interactions within a system along with the nature of the tasks the practicing systems analyst will have to do in order to carry out his job. This chapter describes some additional aids to help a person who is about to enter the systems field or for one who is to become deeply involved with systems analysis from an operations point of view.

Job Description and Expected Duties

The job description of a business systems analyst within one organization contains the following major duties:

1. confers with operating personnel to obtain a clear understanding of the problem and type of data to be processed.
2. prepares definition of problem.
3. devises system requirements and develops procedures.
4. analyzes problem in terms of equipment capability.
5. devises data verification methods.
6. establishes standards for preparation of operating instructions.

Observe that those duties are listed in such a way that you cannot tell if they relate to manual or automated methods. Figure 14-1 shows a much more complete job description, with implications that the analyst will be using a computer where appropriate in system operations.

Those particular activities all seem to be at a reasonably high problem-solving level. Contrast them with what a survey of Systems Departments revealed as the duties of some practicing analysts:

1. supervise the company library.
2. keep time records on labor.
3. handle travel arrangements for company employees.
4. train secretaries and clerical people.
5. provide a lock and key service to assure proper control over entry into certain areas.
6. schedule the use of conference rooms.

Obviously, some Systems Departments around the country have become a dumping grounds for jobs and activities which other

Figure 14.1. Job description of a systems analyst.

1. Upon receiving an assignment from the supervisor of Systems Analysis, visits the departments involved to get a broad view of the problem. By using the standard "pre-study checklist," obtains answers to general questions. Tries to pinpoint specific problems in present systems or determines what the problem is in a completely new system.
2. Reviews findings with supervisor, who determines if there is sufficient reason to move ahead. If there is, obtains specific objectives or works with management to determine them.
3. Gains more precise knowledge by the proper combination of reviewing flowcharts and manuals, interviewing people, and observing workers in the affected areas. Gathers samples of data and develops statistics of work volumes.
4. Generally determines if changes can be made by slightly altering system or if a complete redesign is needed.
5. Prepares possible solutions. Reviews with supervisor and affected departments to ensure whether on a reasonable track.
6. Designs a detailed solution, including:
   a. forms and report layouts.
   b. files, paper, cards, tape, and disk records.
   c. benefit and cost analysis.
   d. controls.
   e. proposal to present to management.
   f. installation schedule.
   g. list of possible disadvantages.
7. Presents complete proposal to management.
8. Turns portions to be mechanized to Programming or Machines Divisions for the detailed work.
9. Completes the documentation file.
10. Helps in installation.
11. Helps Audit Section as needed in its post-installation audit.

people and departments are anxious to get rid of. It is desirable that most of the systems work that you will ever do would be on a much higher plane than the six items mentioned in the paragraph immediately above. Certainly all those activities have to be performed, and by doing some of them you might learn something worthwhile about the company. Hopefully, management will not think of systems in that way as a career, and they would want to move you through that type of work quickly.

Another situation that you must be aware of is that many people with the title of systems analyst may spend a great portion of their time on computer programming as opposed to pure systems work. This has come about because systems work is generally considered to be a higher level position, and people naturally want to move into it. In order to still get the necessary programming done, companies have had to require people to do some of both before completely splitting off the best employees into systems.

Background for Systems Work

Perhaps an ideal background for systems work would be to have several college degrees (in business, engineering, philosophy, and psychology), to be an accomplished speaker and writer, and to have as much as twenty years' experience operating the system for which you are now going to be doing design work. But if requirements anywhere near those were the case, there would be few people who would ever become qualified. Most people who have been chosen have had perhaps only one or a portion of one of those qualifications. Somewhere in between lies a reasonable area for individual selection.

Some companies have not put any particular educational requirements into their selection criteria. At one time, a person was almost required to have a mathematics degree to be a programmer because of the inherent difficulty of programming. But with programming languages being as advanced as they are today and continuing to be simplified for most applications, that is no longer a prerequisite except where it is obviously a mathematics program. Even the person who writes a regular business program and comes to a difficult mathematical problem may be able to get a mathematician to express the solution in a formula, and the programmer can continue from there. Furthermore, many systems analysts are doing their jobs in such a way that they don't have to be accomplished programmers, so this further eliminates the need for a mathematics background to become an analyst.

It is obvious this book has approached systems analysis in such a way that mechanization has not been emphasized. Regardless of how much automation is used to perform certain steps, an amount of pure systems work must be performed. But where computers are used to a high extent, or where computers and not people nor information are considered to be the system, then systems analysts should know a great deal about computers. The best way to gain this knowledge is to be a programmer for a year or so.

But both programming and systems work require a person with a logical mind, and a person who is a good mathematician no doubt thinks in a logical pattern as required. Therefore, the mathematician is often chosen over the others because he would seem to have at least one of the vital qualities that is needed to succeed.

Certainly a business degree would be helpful. If a person has had several courses in accounting and other related subjects, he is in a position to understand some of the basic ways in which business operates and is able to anticipate some of management's requirements. But no amount of college or "book learning" will give a person the equivalent of several years practical experience. A student who has had 12 credits of business courses may have spent 540 hours (12 credits $\times$ 3 hours per week per credit $\times$ 15 weeks per semester) on those subjects. That many hours in course work is equal to only 13.5 (540/40) weeks on the job. But the usual purpose of a college education is not to teach you a series of facts which you will later spew out, but to teach you to think; to go about your affairs in a more organized way; and to learn to get along with the others. So you must recognize that solid business experience may take the place of some college credits in order to prepare you for this type of work, too.

From the above, you might observe that any college background may theoretically prepare a person to do a good job in this field, and as a result you can find English, history, music, language, and psychology majors who have done well in systems work. A difficult thing to measure about the college graduate is how well he is really able to do a job because his courses may have had all true-false, or multiple-choice tests on a recall basis. He may never have had to work closely with anyone else on other than a classroom controlled basis.

There are actually some situations where the college graduate may not be chosen over the nongraduate for a position of this type. Some people in a position of hiring take the following approach concerning the graduate.

1. He may be overloaded with theory and may not have much in the way of practical ideas.
2. He may be too quantitatively-oriented. He may be far ahead of what the business may be able to use in the near future. Many businessmen have shunned the use of much mathematics and rely upon intuition instead.
3. His eyes may be on the presidency long before he would be able to handle the job.
4. He often expects too high a salary in relation to his present practical value.

5. He is more likely to be a job-hopper because he may be very ambitious and want things to happen much faster than they do. The graduate probably won't have trouble getting another job because he is in so much demand.

So you must be willing to face facts and recognize that some organizations would prefer to hire a nongraduate because they think that type of person will be a better investment for them. A business always takes a risk when it hires a person, and some prefer to promote a person to systems work only after he has worked there a while and proved what he can do.

Insofar as speaking and writing abilities are concerned, a college background should have prepared a person in these areas. But a person shouldn't be so proud to think that several courses in those areas will automatically make him successful. He has to absorb that course work in the proper manner in order to make good use of communication skills on the job.

Despite the qualifications a person may seem to have, as evidenced by college or work experience, some firms place a great deal of faith in aptitude tests. These tests don't normally relate directly to programming or systems knowledge, but more to logical and even abstract reasoning. There has been enough evidence to show a direct correlation between test scores and later success in the field, and many businesses put such emphasis on the test that they just won't hire an applicant if he doesn't attain a high score. Practically speaking, they don't want to invest anything in a person if he can't pass that first hurdle with some indication of success. Then there are some people who have no faith in the test and either don't use it at all or justify with other reasons the selection of a person who has not done well on it.

General Philosophy

The attitude you display will have a great deal to do with how successful you become. Obviously, people aren't going to want to work with you if you are surly, constantly critical of everyone and everything, or tend to do things which undermine everyone's confidence in you. I know a person who was hired as the supervisor of a three-man Systems Department. Because of the approach he took, even when invited out to lunch by the treasurer, he did not manage to put one systems change into effect during eighteen months on the job. Every time he tried something, he ran into a stumbling block. It was not because he never had some worthwhile ideas, but because he irritated people. When management had hired him, they made such a big thing of how lucky they were to get him, and they were not willing to later admit they had made a mistake. Everyone was greatly relieved when the fellow realized he wasn't getting anywhere and finally resigned.

The systems analyst must have a basic understanding of the profit motive and what must be done to attain satisfactory profits. Despite any current feelings that profits are bad, that they don't matter, or that companies are in business for purely social reasons, business has proved that profits must be received in order to continue to attract people, obtain money, and stay in business. Even the so-called "nonprofit" organization must be expected to provide reasonable services at reasonable costs. One must recognize there is not a bottomless hole from which to obtain talent and funds.

The term maximizing profits does not infer that an organization is going to make the most profits it can by any means available. The term means you are going to try to make profits as high as you reasonably can by following good, sound, business practices over the long term, not the short term.

The analyst then must know enough about economics and income statements to know what the net profit effect of a particular situation will be. Certainly every organization has some aspects that aren't profitable and never will be, and they have others that by employing better methods could return more, and these must be sought out and improved.

An approach you might follow here is to closely review the goals of the business. This will give somewhat of a guide as to the job you are expected to do. If your management wants a 25 percent return, when the best competitor gets only 18 percent, you know you and everyone else are going to have to work a lot harder. If management's goals are not compatible with your own thinking, then you must determine if you can afford to work in that environment.

Questions

1. Why would there be such a range of duties in the systems field, as shown on the first page of this chapter?

2. Why do many systems analysts spend a lot of their time writing computer programs? Does a systems analyst need to know a lot about programming? Why?

3. Why was a mathematics degree once a prime requisite for becoming a programmer? Why is that not so much the case anymore? Why might a mathematician make a good systems analyst?

4. Does a programmer's aptitude test measure a person's programming experience? Why?

5. What is the difference between systems analysis and systems design? Which is performed first? Always? Which is probably more difficult to do?

6. If there were no Systems Departments at all, would systems work get done? If not, why? If so, how?

7. For each of the following items which appears on an income statement, indicate a specific systems change which might improve that item's contribution to net profit. If there are no systems changes which could improve an item, clearly indicate which ones they are and why.

Sales	Sales Returns
Direct Materials	Direct Labor
Overhead	Depreciation
Selling Expense	G & A Expense
Interest Expense	Income Taxes

CHAPTER 15

Additional Exercises and Case Studies

1. The A Company has a manufacturing operation composed of (1) a process which makes finished parts and (2) a second process which assembles the finished parts into completed units. Because of the manner in which sales orders are received from customers, the level of the work force is constantly altered to the number needed for the backlog of work that is to be done.

The finished parts are standard items which could go into any customer's completed units. It is the manner in which manufacturing process number two takes place that satisfies the requirements of each individual customer.

The following figures indicate the pay scale and other costs to maintain its production work force.

(a) The regular salary is $200 a week. Overtime is paid at a rate of one and a half times the regular pay. But the company has a policy against the use of overtime.
(b) It costs about $300 to hire a person. This includes the costs of advertising, interviewing, medical exams, and a brief orientation period.
(c) It costs $500 to fire a person outright. Most of this is a two-week separation allowance required by the union contract.
(d) A worker gets 80 percent of his regular pay while he is temporarily laid off. It does not cost anything to call back people who have been laid off.

During 1971, fluctuations in the work force at A Company were as follows.

	Process 1	Process 2
Work force on 1/1/71	300	500
Hired on 1/15	50	
Hired on 2/1		75
Laid off on 2/15	50	
Called back on 3/1	50	
Fired on 6/1	25	50
Hired on 8/1	50	25
Laid off on 12/15	25	40

(1) Give at least two possible solutions to the problem surrounding the need to raise and lower the work force so often.
(2) Show what your proposals would cost and what the benefits would be. Make reasonable assumptions and state what they are.

2. Figure 15-1 shows a PERT network that has already been prepared. The estimated times are shown on the activity lines. Following is a list of the activities which can be expedited, the maximum number of days that each can be expedited, and the cost per day to expedite.

Activity	Number of Days	Cost
AE	2	$150
BJ	5	150
BK	1	400
BF	2	25
CJ	2	125
EH	10	140

Figure 15-1.   PERT diagram.

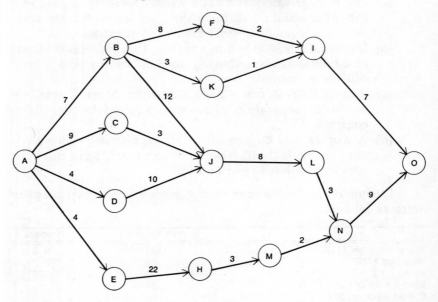

After reviewing the network and the list above, answer the following questions.

(a) What is the critical path?
(b) What would be the minimum additional cost to reduce one day from the schedule to complete the project? What activity is that?

(c) What would be the minimum additional cost to take three days out of the schedule? What would the critical path then be if those three days were taken out?

(d) As the project stands at the moment, could five days be taken out of the schedule? Why?

(e) According to the list above, only one day can be expedited on activity BK at $400. Is there any point in trying to investigate to see if an additional day could be expedited on that activity? Why?

(f) Based upon the beginning information for this problem, can you tell what the estimated cost of the total project is? Be very precise in explaining.

3. Each quarter of the year, an employer must furnish a report to the Social Security Administration. The report must show each employee's name, social security number, and taxable earnings.

At the completion of a year, the employer must furnish the following on each employee to the Internal Revenue Service: name, address, Social Security number, federal income tax withheld, wages subject to income tax, Social Security tax withheld, and total wages subject to Social Security.

(a) From the standpoint of the U.S. Government, what would be the best form in which they might receive the information from employers.

(b) Briefly indicate what might be done to eliminate some of the duplication which is now taking place. Do not fail to include what some of the problems might be with implementing your proposal.

(c) Why do you suppose the duplication of reporting has not already been eliminated?

4. The B Company has 2000 employees. It has a computer which it operates twenty-four hours a day. Most of the applications run are in the category of inventory control, production scheduling, and fixed asset analysis. When the company computerized these applications, profits were quickly increased because new methods of control over the functions could be exercised. You might say these are "money making" applications.

The computer is also used about two hours a day on payroll. The previous manual method of performing payroll was a good one, and no systems changes were made when it was computerized.

There are no important management decisions made as a result of getting payroll information from the computer. This application is just breaking even costwise in relation to the former manual method.

At this point, two major items come to your attention. First, the salesman from a computer service bureau makes a bid to you that he will do, at his location, all the payroll you are now doing on your own computer for $0.32 per employee per week. He is willing to sign a long-term contract on those terms. Second, several other applications have been brought to your attention. One of your analysts has determined that you could save $10,000 annually by computerizing the scrap control program. Another analyst has made a study of another area and believes that $100,000 can be taken out of inventory by better sales forecasting. And a third item is that a neighboring company would like to use the special features of your computer one hour a day on a rental basis. They are willing to pay $75 an hour.

Estimates are that the three possible new applications would require about two hours a day, just equal to the time now spent on payroll.

While it may not be easy to go to management and tell them you want to shift the payroll to someone else and add other work in its place, a systems analyst should be willing to look further into situations of this type. Try to develop reasonable answers to the following questions.

(a) About how much would the company make or lose annually if it made the switch and if all estimates are fairly accurate?

(b) Instead of doing as proposed above, how practical would it be to retain the payroll processing and obtain a second computer in order to take on the three new jobs?

(c) Suppose you prepare a proposal to management on the idea of sending out the payroll. What will you do if you are told it is against company policy to have the payroll processed outside the company?

(d) For what reason might the company have had a policy as stated in (c) above?

(e) What major problems might result from using the outside service bureau? What might be done to solve each of those problems?

(f) What problems might result from renting out time to the neighboring company? How might each of those problems be solved?

5. The major purpose of a budget is to prepare a financial plan which is to be used to guide the business for some future period of time. As time passes, actual results are compared to the budget, and significant differences are isolated and explained.

Following are the monthly sales budget figures (in millions) for 1971 as prepared by the C Company in October 1970.

January	20	July	25
February	21	August	20
March	22	September	21
April	24	October	24
May	24	November	23
June	25	December	22

During 1971, the Financial Department prepared a report which listed budget against actual. Instead of preparing the report on a month-to-month basis, it was shown in a year-to-date style. The January budget was compared to January actual. But in February, the revised year-to-date budget was composed of actual for January and the estimate for February. That revised budget was then compared to year-to-date actual. The following is a summary of the information appearing on the twelve budget comparisons for the year.

Month	Revised Budget	Year-to-Date Actual	Difference
January	20	19	1
February	40	38	2
March	60	58	2
April	82	79	3
May	103	102	1
June	127	125	2
July	150	147	3
August	167	168	(1)
September	189	188	1
October	212	210	2
November	233	233	0
December	255	254	1

Analysts noted the difference between the revised budget and the actual was never more than $3 million at any time and was less than 0.5 percent for the whole year. The company concluded that its budgeting had been pretty accurate and there was no need to explain any differences. Do you agree? Why? If you do agree, justify your position. If not, suggest what should be done.

6. Officials of the D Company are taking a look at the reports they have been receiving on their far-flung branch operations. One of the purposes for having the reports is to be able to determine annual bonuses for the branch managers. Formats of the various reports are shown below. Indicate those which you believe do a very good job and those which may not be a good aid for the purpose of determining manager performance. Explain why you have chosen as you did. (Note: All reports listed with the "best" performance first.)

Report No. 1   Sales per branch (in 000's)

Branch No.	Sales ($)
17	12,000
4	11,276
8	11,143
3	10,004

Report No. 2   Net profit contribution per branch (in 000's)

Branch No.	Net Profit ($)
12	125
26	118
7	102
19	99

Report No. 3   Variation from budget (in 000's)

Branch No.	Variation ($)
24	12
9	17
18	24
23	54

Report No. 4   Variation from budget (%)

Branch No.	Variation (%)
14	1
22	2
27	3
13	6

Report No. 5   Sales per employee

Branch No.	Amount ($)
1	44,000
25	43,215
15	41,109
10	39,076

Report No. 6   Profit per employee

Branch No.	Amount ($)
11	4,835
16	4,612
28	4,498
6	4,100

Report No. 7   Fixed asset investment per employee

Branch No.	Amount ($)
21	36,000
5	34,107
2	33,882
20	32,904

7. Figure 15-2 shows a systems flow chart of a payroll operation. Based upon all the facts shown and implied in the flow chart, determine the minimum information that was contained in each file. (Do not attempt to determine how efficient the operation is.)

8. In accounting operations, a chart of accounts is a list of the names of accounts that will be used to accumulate the dollar effect of transactions. Because of traditional accounting needs, the list is maintained in a specific order other than alphabetic. In fact, a number is usually assigned to each account. When the accounts are in numeric order, they are also in the order required for the most common use.

Below is a list of twenty common accounts shown in alphabetic order. Assign a number to each account so they would be in good order when displayed numerically. Beside each account is a code which indicates relative position on the balance sheet.

CA = Current Asset   CL = Current Liability
FA = Fixed Asset   LT = Long Term Debts
OA = Other Asset   NW = Net Worth

Figure 15-2.   Flow chart.

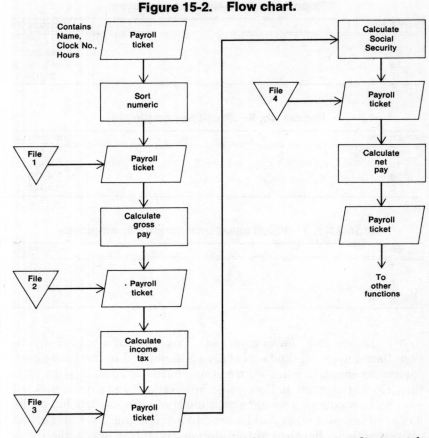

Design the codes in such a way that it is easy to distinguish among the six types of balance sheet accounts. A three-digit code is of sufficient length.

Accounts Payable (CL)	Current Portion—Mortgage (CL)
Accounts Receivable (CA)	Goodwill (OA)
Accrued Payroll (CL)	Inventory (CA)
Accrued Taxes (CL)	Investment in Subsidiary (OA)
Accumulated Deprec. (FA)	Mortgage Payable (LT)
Allowance for Bad Debts (CA)	Preferred Stock (NW)
Cash in Bank—1st Nat'l (CA)	Prepaid Expenses (CA)
Cash in Bank—2nd Nat'l (CA)	Property, Plant, & Equipment (FA)
Cash on Hand (CA)	Retained Earnings (NW)
Common Stock (NW)	Treasury Stock (NW)

9. Many businesses take a complete count of their physical inventory each year. During the year they may maintain detailed inventory records of sales, receipts, and current balances. After taking the inventory, they compare the physical count to the book count and make adjusting entries to bring the books up to date. The following narratives describe how four different companies take the physical counts.

E Company

At inventory time, employees of this company make up the following column headings on sheets of paper: name of item, quantity counted, unit cost, and extended amount. Each sheet is long enough to list twenty-five items; enough sheets are reproduced to contain all the items the company has.

An employee takes the sheets and proceeds to the inventory area. For each item in stock, he records the item name and then counts the quantity on hand and writes it in the proper place. When all items have been recorded in this fashion, the sheets are taken to the Accounting Department.

Employees look up from file records what the unit cost is of each item. These values are placed on the appropriate lines of the inventory sheets. Then a desk calculator or comptometer is used to multiply the quantity by the unit cost, and the extended amount is written on the sheet. The extended amounts are added to arrive at an inventory value.

The inventory value thus calculated is compared to the figure in the accounting records. If the book figure is greater, the difference is subtracted from the books. If the book figure is smaller, the difference is added to the books.

F Company

About two weeks before the inventory is to be taken, employees go through appropriate inventory and financial records and then prepare forms in the following format.

Name of Item	Unit Cost ($)	Quantity	Extension
Bolts	.12		
Braces	.29		
Hinges	1.45		
Nuts	.02		

The inventory is eventually counted, and since the unit cost of each item was put on the sheets beforehand, the multiplication by quantity can take place without the need to go back to file records. The total inventory value is eventually related to an accounting record for the purpose of making an entry to the books.

G Company

This company has all its inventory records in computer processable form (punched cards, magnetic tape, or disk file). The night before inventory taking, the computer punches out a card for each inventory item, the card containing part number, description, bin location, unit cost, and quantity on hand (book quantity) per computer records. The cards are interpreted so people can readily identify them. They are then sorted to bin location and placed on the shelf next to the stock.

The people taking inventory count how many there are for an item, check to make sure they have the right card, and write the quantity in a special place on the card. The cards are all collected, the quantities keypunched into the cards, and the quantities multiplied times the unit cost. The cards are then sorted to part number and a report run off in the following format:

Part No.	Description	Unit Cost ($)	Book Quantity	Physical Count	Difference	Dollar Value Per Count
17	Bolts	.12	150	144	6	17.28
26	Braces	.29	9	9	—	2.61
165	Hinges	1.45	74	50	24	72.50
873	Nuts	.02	2004	2000	4	40.00

On those items that have an unusually high "difference," an employee goes to the bin area and counts again to see if the physical count may have been off. If so, an adjustment is made to the card and the report.

Since the book figures and physical count are now both in computer processable form, the computer can adjust the book figures to what was actually on the shelf.

H Company

This company uses an approach substantially different from all the others. It has all beginning records in computer form: several weeks before time to take the inventory, it sorts all items to bin location and prints two copies of a form that looks like this:

Bin Location	Part Number	Description	Count
14	44	13" Wheels	
14	176	14" Wheels	
14	283	15" Wheels	
14	971	16" Wheels	
15	412	LF Fenders	

On inventory day, all receipts and shipments are suspended. One copy of the form is placed near the material, and counts are made and recorded in the morning. Then all inventory forms are gathered. In the afternoon, the process is repeated with the second copy of the form by a different group of people. The forms are eventually sent to the Data Processing Department where the necessary information is keypunched for both morning and afternoon counts. Once the counts are in computer form, an analysis like the following is run off.

Bin Location	Part Number	Description	First Count	Second Count	Difference
14	44	13" Wheels	42	42	—
14	176	14" Wheels	35	35	—
14	283	15" Wheels	127	128	1
14	971	16" Wheels	4	4	—
15	412	LF Fenders	8	8	—

On every item that shows a difference, a special group of people is sent to make a third count and to reconcile the difference that exists. When these problems of the basic count are cleared up, a report is run which shows the difference between physical and book, and adjusting entries are made.

Answer the following questions with respect to the four different methods of inventory taking described above.

(a) Give a strong advantage of each method.
(b) Give a major disadvantage or flaw of each method.
(c) Indicate the general type of business which might make good use of each type described.
(d) Rank the methods according to the following criteria.

Feature	Highest	Lowest
(1) Cost (relative to number of items)		
(2) Accuracy		
(3) Disruption to regular warehouse activities		
(4) Span of time required to complete		
(5) Confidentiality of information		

(e) Make a critical analysis of the following summary. Its purpose is to report the difference between physical and book figures.

Items Which Had a Difference	Amount of Difference ($)
137	1,426

10. For many years, the I Company used a form such as that shown in Figure 15-3 to keep track of what their individual customers owed them. When any activity occurred in the account, posting took place as indicated, and an up-to-date balance was always shown. The form had enough posting room so that quite a bit of the historical activity with the customer was available.

Figure 15-3.   Accounts receivable ledger card.

CUSTOMER # AND NAME 12794 Doe Products				ADDRESS & PHONE 1 Main St. 442-1300 Anytown, USA 00000			
Sales		Items Paid	Payments		Adjustments		Balance
Date	Amount		Date	Amount	Amount	Ref.	−0−
2/9/71	12,000	✓					12,000
2/14/71	4,400	✓					16,400
			3/8/71	12,000			4,400
3/18/71	1,427	✓					5,827
					−127	V 46	5,700
			3/15/71	4,400			1,300
4/17/71	9,000	✓					10,300
			4/17/71	1,300			9,000
4/18/71	2,000	✓					11,000
4/22/71	3,500						14,500
			5/17/71	11,000			3,500

Then the company began considering the use of a computer to keep track of its accounts receivable. The computer specialists soon realized that they would not be able to print as much historical detail each computer cycle as the previous ledger card showed, so their proposal for a printout as a replacement for the ledger card was as shown in Figure 15-4.

Figure 15-4. Accounts receivable listing for May 1971.

Customer #	Customer Name	Old Balance	Sales	Payments	Adj.	New Balance
12794	DOE PRODUCTS	14,500		11,000		3,500
97463	Zale Toys	4,200	1,876	3,200	+185	3,061

Answer the following questions.

(a) Why would accounts receivable personnel want to see records in detail such as that shown in Figure 15-3.
(b) Would there be a lesser need to review the records (for the reasons in a above) if all the data was eventually to be printed by a computer?
(c) What will probably be the reaction of accounts receivable personnel to the proposal by the computer people?
(d) Suggest a sound operating method whereby the computer can do the posting in approximately the same detail shown in Figure 15-4 but where it is done in such a way that the basic purposes of accounts receivable personnel are served.
(e) Why might a company want to switch such record keeping to a computer?
(f) Why can't the computer print the report in the same way as it was done previously at a lower cost?

11. At the time it was in need of money to finance certain expansion programs, the J Company negotiated a special type of loan with a bank on the following basis. The bank would loan the money for a whole year at an interest rate of 8 percent. No pay-

ments on principal were required during the year, but interest payments had to be made monthly. The company agreed to keep at least 20 percent of the amount of the loan in its checking account at the bank throughout the year. On January 2, the company estimated that it needed $1 million for its expansion purposes, and it borrowed the necessary money.

(a) How much did they have to borrow, assuming they had nothing in their checking account?
(b) How much interest did they have to pay for the year?
(c) What was the effective rate of interest on the loan (what they really had use of in relation to the interest charge)?
(d) Suppose their real need for the funds were such that they could have borrowed and paid back in the following manner.

Date	Borrowed ($)	Paid Back ($)
1/2	1,000,000	
3/31		600,000
5/1	200,000	
9/1	100,000	
10/31		300,000
12/1	200,000	

What would have been their total interest charge for the year if they would have borrowed the exact amounts shown at 9 percent?
(e) Why would a bank require that a company keep 20 percent of its loaned amount in a checking account?

12. The promoters of a major sports car race were interested in good control over cash receipts and related activities at a three-day race. The company had good physical control through high fences, etc. It was decided to charge $2 for Friday only, $4 for Saturday only, $8 for Sunday only, and $12 for all three days. It was felt that fans should have unlimited checkout privileges.

Design a brief system that would adequately control people so the promoters wouldn't lose a lot of money due to poor collection and crowd control.

13. In an effort to get a better understanding of what its inventory situation was like, the K Company got the following listing for a sample of its products. (The listing contains what is felt to be a very representative sample of what the real situation is.)

Part No.	Quantity on Hand	Unit Cost ($)	Sales This Year	Sales This Month
12	140	36.00	560	80
87	5	512.00	12	2
144	3,120	1.20	14,000	970
891	18	162.00	15	1
1007	400	4.27	8,200	1,000
1218	174	9.00	1,763	512
1473	18	12.00	300	41
1680	960	2.70	4,320	318
1909	250	7.20	1,820	143

(a) Based on the ending inventory, what is the turnover for the year?
(b) Calculate the month's supply on hand for each item. Which items have far too much stock on hand? State the criteria you have chosen to use as a standard amount.
(c) What types of things can be done to bring inventory back in line? What are some of the consequences of doing that? Be specific.

14. The L Company is known for doing many things incorrectly. Its income statement, shown in Figure 15-5, contains many items that are not what they would otherwise have been if it had not been for operating mistakes. Some of the penalties they incurred during the year are as follows:

(a) Half of their raw material purchases were made on terms of 2 percent, 10 days, net 30. (They would have gotten a discount if they had paid the bill within 10 days.) They missed the discount about half of the time.
(b) The company incurred penalties as a result of delivering certain products to its customers after a specified due date. These penalties amounted to 0.5 percent of sales.
(c) The company had about $1 million in extra cash all year long. Interest rates averaged 7 percent.
(d) During the year, .75 percent of raw material on hand either spoiled or became obsolete due to being kept too long.
(e) Production lines were temporarily shut down 2 percent of the time because materials weren't available. Workers were paid during this time.
(f) The company operated its equipment at about 90 percent of reasonable capacity.
(g) Income taxes were paid too late, and a 10 percent penalty was assessed.

Answer the following questions with respect to operations of the L Company.

(1) Calculate the approximate amount of each of the seven expenses the company incurred over the normal expenses.
(2) For each of the expenses, show the item on the income statement where the amount is most likely buried. (A company's books may reflect such problems, but a published financial statement usually does not.)
(3) Give a sound reason for what might have happened within the company to cause each of the situations. Then give a very brief idea as to what might have been done to clear up that situation so it does not occur again.

Figure 15-5.   L Company statement of income for the year ending December 31, 1970.

Sales		$10,000,000
Cost of Goods Sold:		
Beginning Inventory	$ 300,000	
Raw Material Purchases	4,600,000	
Total Available	4,900,000	
Less: Ending Inventory	500,000	4,400,000
Gross Profit		5,600,000
Expenses:		
Direct Labor	3,000,000	
Depreciation	400,000	
Administration and Selling	1,000,000	
Interest	150,000	4,550,000
Net Income Before Taxes		1,050,000
Income Taxes		525,000
Net Income		$ 525,000

15. In 1970, Jones had total sales of $10 million of which 28 percent were for cash. His accounts receivable were $700,000 on 12/31/69 and $900,000 on 12/31/70. Calculate his accounts receivable turnover for 1970.

Suppose Jones' store was open for business every day of the year. About how many days of charge sales did his accounts receivable represent on 12/31/70? (Note: Use the same concept as month's supply on hand in inventory.) Under what circumstances would that number be acceptable? Under what circumstances would it be much too high?

16. Name two common ways of precoding sales transactions so that time is saved and errors reduced at the point of sale.

17. When a major league baseball club is about to enter the World Series, it prints tickets and makes sales (all mail order) for all seven games. Quite often the series does not go the seven games, and refunds have to be made.

Should the club cash all customer checks as they are received and then make out their own checks as refunds when and if that becomes necessary? Or should they hold all checks to see if a particular game will be played, and if it isn't, merely mail the customer's check back to him?

18. Figure 15-6 shows four different ways of copying a disk file. The process is often called dumping, even though it does not destroy the data on the original disk. Describe a situation where each approach might be used to greatest advantage.

19. One of the major purposes behind controls is to try to prevent or catch fraud. List three other major purposes of controls.

20. Name a situation where it would be very desirable for a computer programmer to run a program that he wrote.

21. A "bond" is a special type of insurance policy that a company may obtain on an employee to recover if the person should embezzle from the company. If all people responsible for handling money are bonded, there is less reason to institute tight controls since the company can't lose anyway. True or false? Why?

22. In an attempt to put inventory items into classes relative to both their dollar value and activity, some companies establish A, B, and C categories. The purpose of this is to then apply different rules for replenishing items according to their category. Assume the rules as you obtain them for a company are as follows:

(a) "C" items are to be ordered in a one-year supply unless the first digit is a "one," in which case you order only a six-month supply. But if the part number has an "X" suffix, that is the designation for a discontinued item.
(b) On any "B" items a one-month supply is to be ordered.
(c) Several things determine the reorder quantity that is to apply on "A" items. If shipping code is an "A" (meaning air freight), order a projected one-week supply. If the item is bought from a subsidiary, refer the order to the department manager. On all other items, order the amount actually consumed in the previous calendar week. If that number should be zero, order 2 percent of the actual usage for last year.

Figure 15-6. Four different ways of copying a disk file.

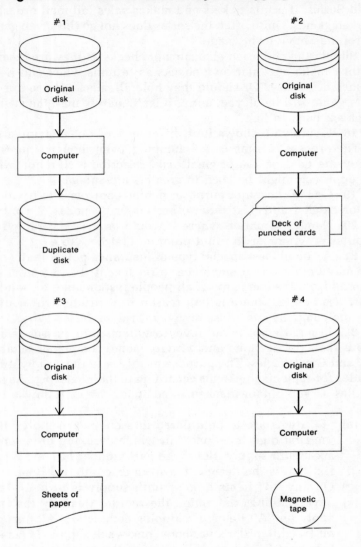

Prepare a decision table that would show the above rules for reordering.

23. Figure 15-7 shows the decision table that an analyst prepared for the shipping of orders to customers. What obvious error was made? Suggest a reasonable way to correct it.

Figure 15-7.   Decision table.

	1	2	3	4	5
Is it an urgent shipment?	N	N	N	N	Y
Is weight 10–100 lb?	Y	N	N	N	—
Is weight 101–5000 lb?	—	Y	N	N	—
Is weight 5001–20,000 lb?	—	—	Y	N	—
Is weight over 20,000 lb?	—	—	—	Y	—
Use U.S. Post Office	X				
Use Railway Express		X			
Use truck freight			X		
Use rail freight				X	
Use air freight					X

24. The MO Division of a hypothetical company deals strictly on a "mail order" basis. People can mail, phone, or deliver their orders to the firm's various locations, and customers can pick up the goods or have the items mailed to them.

The company does not keep any record of "back orders." That is, when they do not have an item to satisfy a request for it, they do not keep a record of the order to fill it when the item is available. Instead, they just inform the customer the item is temporarily out of stock, and that he may place a new order for it in ten days. Over the years, the company has made spot checks and found that the customer has ordered again in 75 percent of the cases.

Now the company wants to take another look at the policy to see if it should be maintained or changed. Based upon their own cost estimates and what they can learn from other stores in the same line of business, they estimate it would cost them about $0.75 to handle the average backorder. This figure would cover the necessary clerical help, forms, and filing space.

They average about 1000 backorders a day throughout their organization. The company processed 6,000,000 orders in 1970.

Based upon the summary income statement shown below, about how much difference in net income before taxes would there have been if the company had processed back orders?

Summary income statement MO division for the year ending December 31, 1970	
Sales	$150,000,000
Cost of Goods Sold	75,000,000
Gross Profit	$ 75,000,000
Administrative & Selling Expenses	45,000,000
Net Income Before Taxes	$ 30,000,000

APPENDIX

Suggested Literature for
the Practicing Systems Analyst

Books

An Annotated Bibliography for the Systems Professional, 2nd ed., 1970. Association for Systems Management, 24587 Bagley Road, Cleveland, Ohio 44135.

Data Communications in Business; An Introduction, American Telephone & Telegraph Company, New York, 1965.

Ideas for Management. Proceedings of the International Systems Meeting of the Association for Systems Management, 24587 Bagley Road, Cleveland, Ohio 44135.

Nadler, Gerald, *Work System Design: The Ideals Concept.* Homewood, Ill.: Richard D. Irwin, Inc., 1967.

Neuschel, Richard F., *Management by System.* New York: McGraw-Hill Book Company, Inc., 1960.

Office Clerical Time Standards. Association for Systems Management, 24587 Bagley Road, Cleveland, Ohio 44135.

Periodicals

Data Processing Digest. 1140 South Robertson Boulevard, Los Angeles, Calif. 90035. Articles are brief enough to save you time, but clear enough to steer you to what interests you most.

Datamation. F. D. Thomson Publications, Inc., 141 East 44th Street, New York, N.Y. 10017.

Journal of Systems Management. Association for Systems Management, 24587 Bagley Road, Cleveland, Ohio 44135. A monthly section "Worth Reading" helps pinpoint current literature which should help you with specific problems.

INDEX

257

71 72 73 74   9 8 7 6 5 4 3 2 1